Desirée Joosten-ten Brinke · Mart Laanpere (Eds.)

Technology Enhanced Assessment

19th International Conference, TEA 2016
Tallinn, Estonia, October 5–6, 2016
Revised Selected Papers

Editors
Desirée Joosten-ten Brinke
Fontys University of Applied Sciences
Welten Institute Open University
Heerlen
The Netherlands

Mart Laanpere
Tallinn University
Tallinn
Estonia

ISSN 1865-0929   ISSN 1865-0937   (electronic)
Communications in Computer and Information Science
ISBN 978-3-319-57743-2   ISBN 978-3-319-57744-9   (eBook)
DOI 10.1007/978-3-319-57744-9

Library of Congress Control Number: 2017940626

© Springer International Publishing AG 2017
This work is subject to copyright. All rights are reserved by the Publisher, whether the whole or part of the material is concerned, specifically the rights of translation, reprinting, reuse of illustrations, recitation, broadcasting, reproduction on microfilms or in any other physical way, and transmission or information storage and retrieval, electronic adaptation, computer software, or by similar or dissimilar methodology now known or hereafter developed.
The use of general descriptive names, registered names, trademarks, service marks, etc. in this publication does not imply, even in the absence of a specific statement, that such names are exempt from the relevant protective laws and regulations and therefore free for general use.
The publisher, the authors and the editors are safe to assume that the advice and information in this book are believed to be true and accurate at the date of publication. Neither the publisher nor the authors or the editors give a warranty, express or implied, with respect to the material contained herein or for any errors or omissions that may have been made. The publisher remains neutral with regard to jurisdictional claims in published maps and institutional affiliations.

Printed on acid-free paper

This Springer imprint is published by Springer Nature
The registered company is Springer International Publishing AG
The registered company address is: Gewerbestrasse 11, 6330 Cham, Switzerland

Communications
in Computer and Information Science   653

Commenced Publication in 2007
Founding and Former Series Editors:
Alfredo Cuzzocrea, Dominik Ślęzak, and Xiaokang Yang

Editorial Board

Simone Diniz Junqueira Barbosa
  *Pontifical Catholic University of Rio de Janeiro (PUC-Rio),
  Rio de Janeiro, Brazil*
Phoebe Chen
  La Trobe University, Melbourne, Australia
Xiaoyong Du
  Renmin University of China, Beijing, China
Joaquim Filipe
  Polytechnic Institute of Setúbal, Setúbal, Portugal
Orhun Kara
  TÜBİTAK BİLGEM and Middle East Technical University, Ankara, Turkey
Igor Kotenko
  *St. Petersburg Institute for Informatics and Automation of the Russian
  Academy of Sciences, St. Petersburg, Russia*
Ting Liu
  Harbin Institute of Technology (HIT), Harbin, China
Krishna M. Sivalingam
  Indian Institute of Technology Madras, Chennai, India
Takashi Washio
  Osaka University, Osaka, Japan

More information about this series at http://www.springer.com/series/7899

Preface

The objective of the International Technology-Enhanced Assessment Conference (TEA) is to bring together researchers and practitioners with great ideas and research on this important topic. This volume of conference proceedings provides an opportunity for readers to engage with refereed research papers that were presented during the 19th edition of this conference, which took place in Tallinn, Estonia, at Tallinn University. Each paper was reviewed by at least three experts and the authors revised their papers based on these comments and discussions during the conference.

In total, 16 submissions of 40 authors were selected to be published in this volume. These publications show interesting examples of current developments in technology-enhanced assessment research. The increasing use of technology takes place in summative and formative assessment contexts in all different domains. We see a progression in research in the measurement of higher-order skills, such as collaborative problem-solving or presentation skills, but also in the development of tools for assessors. Additionally, research that provides guidelines for policy on e-assessment, for example, guidelines for authentication control or guidelines for assessment on personal devices is presented. As massive open online courses (MOOCs) offer a good opportunity for ongoing learning, the role of self-assessment becomes more important. Research on self-assessment in MOOCs is therefore also increasing. The papers will be of interest for educational scientists and practitioners who want to be informed about recent innovations and obtain insights into technology-enhanced assessment. We thank all reviewers, contributing authors, and the sponsoring institutions for their support.

February 2016

Desirée Joosten-ten Brinke
Mart Laanpere

Organization

The 2016 International Technology-Enhanced Assessment Conference was organized by Tallinn University, the Welten Institute – Research Centre for Learning, Teaching and Technology of the Open University of The Netherlands, and the Embedded Assessment Research Group of LIST, the Luxembourg Institute of Science and Technology.

Executive Committee

Conference Chairs

Desirée Joosten-ten Brinke	Open University of The Netherlands and Fontys, University of Applied Sciences, The Netherlands
Mart Laanpere	Tallinn University, Estonia

Local Organizing Committee

Desirée Joosten-ten Brinke	Open University of The Netherlands and Fontys, University of Applied Sciences, The Netherlands
Mart Laanpere	Tallinn University, Estonia
Karin Org	Tallinn University Conference Center, Estonia
Mieke Haemers	Open University of The Netherlands, The Netherlands
Tobias Ley	University of Tallinn, Estonia
Kai Pata	University of Tallinn, Estonia
Terje Väljataga	University of Tallinn, Estonia
Kairit Tammets	University of Tallinn, Estonia
Sille Sepp	University of Tallinn, Estonia

Program Committee

Marco Kalz	Welten Institute, Open University of The Netherlands, The Netherlands
Eric Ras	Luxembourg Institute of Science and Technology, Luxembourg
Davinia Hernández-Leo	Universitat Pompeu Fabra, Barcelona, Spain
Peter Reimann	University of Sydney, Australia
William Warburton	Southampton University, UK
Silvester Draaijer	Vrije Universiteit Amsterdam, The Netherlands
Samuel Greiff	University of Luxembourg, Luxembourg
Mark Gierl	University of Alberta, Canada
Lester Gilbert	University of Southampton, UK
Peter Van Rosmalen	Open University of The Netherlands, The Netherlands
Stefan Trausan-Matu	Politehnica University of Bucharest, Romania

Geoffrey Crisp	RMIT University, Australia
George Moerkerke	Open University of The Netherlands, The Netherlands
Denise Whitelock	Open University, UK
Jeroen Donkers	University of Maastricht, The Netherlands
Dai Griffiths	cetis
Marieke van der Schaaf	University of Utrecht, The Netherlands
Mieke Jaspers	Fontys, University of Applied Sciences, The Netherlands
Desirée Joosten-ten Brinke	Welten Institute, Open University of The Netherlands, The Netherlands
Ellen Rusman	Welten Institute, Open University of The Netherlands, The Netherlands
Annette Peet	SURFnet, The Netherlands
Jorik Arts	Fontys, University of Applied Sciences, The Netherlands
Kelly Meusen-Beekman	Open University of The Netherlands, The Netherlands
Gerry Geitz	Open University of The Netherlands, The Netherlands
Karin Gerritsen-van Leeuwenkamp	Open University of The Netherlands, The Netherlands
Patrick Griffin	University of Melbourne, Australia
Thibaud Latour	Luxembourg Institute of Science and Technology, Luxembourg
Romain Martin	University of Luxembourg, Luxembourg
Beatriz Florian-Gaviria	Universidad del Valle, Colombia

Contents

A First Step Towards Synthesizing Rubrics and Video for the Formative
Assessment of Complex Skills 1
   *Kevin Ackermans, Ellen Rusman, Saskia Brand-Gruwel,
and Marcus Specht*

Case Study Analysis on Collaborative Problem Solving
Using a Tangible Interface 11
   Dimitra Anastasiou and Eric Ras

Feedback Opportunities of Comparative Judgement: An Overview
of Possible Features and Acceptance at Different User Levels. 23
   *Roos Van Gasse, Anneleen Mortier, Maarten Goossens, Jan Vanhoof,
Peter Van Petegem, Peter Vlerick, and Sven De Maeyer*

Student Teachers' Perceptions About an E-portfolio Enriched
with Learning Analytics.. 39
   *Pihel Hunt, Äli Leijen, Gerli Silm, Liina Malva,
and Marieke Van der Schaaf*

Support in Assessment of Prior Learning: Personal or Online? 47
   Desirée Joosten-ten Brinke, Dominique Sluijsmans, and Wim Jochems

The Role of Formative Assessment in a Blended Learning Course 63
   Sharon Klinkenberg

Self- and Automated Assessment in Programming MOOCs 72
   *Marina Lepp, Piret Luik, Tauno Palts, Kaspar Papli, Reelika Suviste,
Merilin Säde, Kaspar Hollo, Vello Vaherpuu, and Eno Tõnisson*

Assuring Authorship and Authentication Across the e-Assessment Process. . . 86
   Ingrid Noguera, Ana-Elena Guerrero-Roldán, and M. Elena Rodríguez

Today - Only TEA and no CAAffee. But Tomorrow? 93
   Rein Prank

A Hybrid Engineering Process for Semi-automatic Item Generation. 105
   Eric Ras, Alexandre Baudet, and Muriel Foulonneau

Assessing Learning Gains 117
   Jekaterina Rogaten, Bart Rienties, and Denise Whitelock

Requirements for E-testing Services in the AfgREN Cloud-Based
E-learning System.. 133
   Salim Saay, Mart Laanpere, and Alex Norta

A Review of Interactive Computer-Based Tasks in Large-Scale Studies:
Can They Guide the Development of an Instrument to Assess Students'
Digital Competence?.. 148
   Leo A. Siiman, Mario Mäeots, and Margus Pedaste

Design and Development of IMS QTI Compliant Lightweight Assessment
Delivery System... 159
   *Vladimir Tomberg, Pjotr Savitski, Pavel Djundik,
   and Vsevolods Berzinsh*

Exploring a Solution for Secured High Stake Tests on Students' Personal
Devices... 171
   Ludo W. Van Meeuwen, Floris P. Verhagen, and Perry J. Den Brok

What Types of Essay Feedback Influence Implementation: Structure Alone
or Structure and Content?.. 181
   *Denise Whitelock, Alison Twiner, John T.E. Richardson, Debora Field,
   and Stephen Pulman*

Author Index 197

A First Step Towards Synthesizing Rubrics and Video for the Formative Assessment of Complex Skills

Kevin Ackermans[✉], Ellen Rusman, Saskia Brand-Gruwel, and Marcus Specht

Welten Institute, Open Universiteit, Heerlen, The Netherlands
{Kevin.Ackermans,Ellen.Rusman,Marcus.Specht,
Saskia.Brand-Gruwel}@ou.nl

Abstract. For learners, it can be difficult to imagine how to perform a complex skill from textual information found in a text-based analytic rubric. In this paper we identify three deficiencies of the text-based analytic rubric for the formative assessment of complex skills. We propose to address the text-based analytic rubric's deficiencies by adding video modeling examples. With the resulting Video Enhanced Rubric we aim to improve the formative assessment of complex skills by fostering learner's mental model development, feedback quality and complex skill mastery.

Keywords: Video · Rubrics · (Formative) assessment · Complex skills · Mental models

1 Introduction

A text-based analytic rubric can be an effective instrument for the formative assessment of complex skills, providing a detailed description of each level of complex skill mastery. This detailed description provides structured and transparent communication of the assessment criteria, providing a uniform assessment method that fosters insight into complex skills acquisition and fosters learning [1]. Apart from being an effective instrument, the rubric is an instrument that can be implemented to address the lack of explicit, substantial and systematic integration of complex skills in the Dutch curriculum [2, 3].

This paper proposes that implementing a rubric for the specific purpose of formatively assessing complex skills presents several deficiencies. For instance, it can be hard for learners to form a rich mental model of a complex skill from a rubric alone. To understand the deficiencies of a rubric when applied for this specific purpose, we need to understand the characteristics of the complex skills we wish to foster. For this paper, the term complex skill is used for the complex skills of presentation, collaboration and information literacy. Also, the term rubric is used for a text-based analytic rubric.

One of the main characteristics of complex skills is that they are hard to learn, requiring an estimated five hundred hours to acquire [4]. Complex skills are skills that are comprised of a set of constituent skills which require conscious processing [5]. A single skill, such as typing, differs in complexity from a complex skill such as giving a presentation. Giving a presentation is comprised out of several constituent skills, such as using presentation software, communication with the audience and the use of information. A rubric does not provide the modeling example needed to contextualize the complex skill and foster a rich mental model [6].

We expect a video modeling example to provide contextualized illustration, supportive- and procedural information to the text-based qualities of a rubric. Bearing practical implementation within Dutch education in mind, using video can provide clear and consistent modeling examples across classrooms.

The highlighted area of Fig. 1 illustrates the core of the general problem definition found in paragraph 8.1 and its position amongst the upcoming theoretical paragraphs.

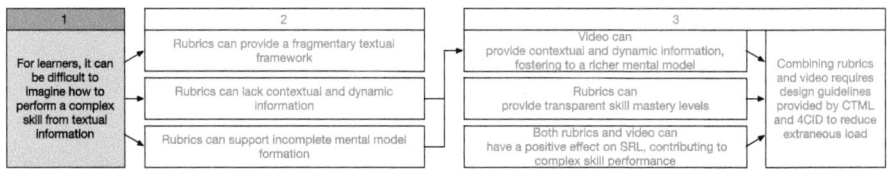

Fig. 1. The core summary of par. 1

2 Specific Problem Definition

Having stated the general problem definition of this paper, a rubric has three specific deficiencies when used for the (formative) assessment of complex skills. In exploring the deficiencies of a rubric, the requirements for video modeling examples present themselves.

Firstly, a rubric provides a fragmentary text-based framework to assess the learner's proficiency of complex skills. Complex skills are comprised of several constituent skills, generally identified by experts through the process of task analysis. However, even though experts are qualified to analyze the constituent skills that form a complex skill, variances may occur in the identification and subsequent hierarchy of constituent skills. Causes for variances can be found in the varying level of expertise and the effect of expert tacit knowledge, causing expert knowledge to be difficult to transfer to a text-based rubric. The varied fragments of a complex skill identified by expert may result in a fragmentary rubric of a complex skill [4]. The fragmentary nature of a rubric may result in an incomplete and moderately reliable assessment of complex skills [7]. An assessment tool with both the ability to assess progress on constituent skill level and assess the progress of the coordination, combination and integration of constituent skills is needed to improve a text-based rubric [5, 6]. We expect videos can 'fill in the gaps' and prevent the fragmentation of constituent skills in a text-based rubric as video provides the learner with the opportunity to personally encode dynamic and contextual information from the video modeling example of the complex skill. Dynamic information refers to information extracted from dynamic stimuli such as video, whereas contextual information refers to information that is connected with real world attributes in order to represent the complex skill within a natural context [8, 9].

Secondly, a rubric is likely to provide an incomplete mental model of a complex skill for low-level learners who perform a Systematic Approach to Problem Solving (SAP) analysison a rubric to achieve a passing grade. In the Four Component Instructional Design

(4C/ID) methodology, performance objectives are formulated to work on constituent skills. The four elements of a performance objective are described on a textual level to form the levels of an analytic rubric. These elements are the (1) tools (2) conditions (3) standards and (4) action the learner should perform to meet the performance objective [4]. Low-level learners use the information found in the performance objectives to strategically reach a passing grade [10, 11]. The learner tackles the problem of reaching a passing grade by analyzing the levels of a rubric for phases and sub-goals, effectively performing a Systematic Approach to Problem Solving (SAP) analysis on the text-based rubric. However, because a learners' mental model is not solely built on the SAP found in a text-based rubric, a rubric should be accompanied with relevant modeling examples. A rich mental model is built on the rich modeling example found in an experts' execution of a complex skill. From the execution of the experts' action, a consciously controlled mental processes can be interpreted by the learner. These expert actions can be visualized in a video modeling example. We expect that the implementation of video modeling combined with a rubric conveys the necessary information to form a rich mental model because the learner may encode the rich mental model found in the actions of the expert performing the task.

Thirdly, the text-based form of a rubric inherently lacks contextual and dynamic information. As complex skills are comprised out of several constituent skills, the priority, sequence and physical performance of the complex skill need to be observed by the learner to supplement the textual assessment criteria with context and dynamic information [9].

In conclusion, we have analyzed several problems with the current implementation of a text-based rubric for the learning process and formative assessment of complex skills. Having specified these problems, we expect the synthesis of video modeling examples and a rubric in the form of a Video Enhanced Rubric (VER) may address these problems, and we move on to the theoretical background of this paper (Fig. 2).

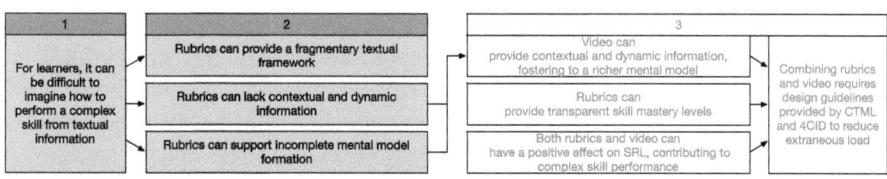

Fig. 2. The core summary of par. 2

3 Theoretical Background of the VER

Having proposed the VER for the (formative) assessment of complex skills, we explore the theoretical arguments for its effectiveness in this paragraph. We start with a brief background on both the text-based rubric and video. Hereafter we analyze overlapping qualities of rubric and video modeling example and finally discuss theory concerning the synthesis of these media into a VER. The theory concerning mental model development, feedback quality, and complex skill mastery is also briefly discussed, as these are the factors we wish to foster by implementing the VER.

As we are using a rubric as a foundation for the VER, we first need to have insight in the qualities of a rubric to understand how they foster the development of complex skills. The transparent grading criteria found in a rubric may be its most important quality, fostering assessment quality and effectiveness, positively influencing the learners' performance [12–17]. From a learner's standpoint, the transparency of a rubric may aid the feedback process by allowing the learner to review the received feedback with the help of the rubric, and provide (self- and peer) feedback based on the rubric. The transparency of a rubric may also allow low-achieving learners to strategically reach a passing grade by providing valuable insight into the minimum requirements per constituent skill [10, 11]. Furthermore, the transparent assessment criteria found in a rubric may reduce anxiety in learners by clarifying teacher expectations if a minimum intervention time of two weeks is factored into the research design as a boundary condition [18–20]. For the assessment of complex skills, a rubric fosters a beneficial effect on self and peer assessment by increasing the validity and reliability of self and peer assessment [1, 7, 21].

One of the research questions of this paper concerns the effect of the VER on the mental model development of a complex skill. A key method used to assess mental model development and also used in this paper, is the creation of a contextual map by the learner. By fostering the representation of knowledge through clear textual descriptions, a rubric fosters the quality of the learners' conceptual maps [1].

Concluding, we expect the transparent skill mastery levels provided by a rubric to foster feedback quality and mental model development of complex skills.

Having summarized the qualities of a rubric and its contribution to the development of complex skills; we proceed to explore the theoretical benefits of the video modeling examples we wish to add to the rubric. We expect better performance of learners on several dimensions making use of the VER, ranging from a richer mental model as a result of the dynamic superiority effect, to increased mastery on specific complex skills.

The dynamic superiority effect proposes advantages of moving images over static visual images [9]. This theory states that moving pictures increase learners' performance because they are remembered better, contain more information, provide more cues to aid retrieval from long-term memory, attract more attention from the learner and increase learner engagement [9]. The dynamic superiority effect also states that the encoding of moving images in a learners' mind involves extraction of semantic and dynamic information, thereby supplementing the rubric's lack of contextual and dynamic information. In the retrieval of the expert modeling example from long-term memory, the modeling example is decoded by the learner. This provides a rich mental model of the complex skill, as the learner visualizes the expert's actions [6, 9]. A richer mental model resulting from the decoding of moving images can relate to higher performance of the learners' complex skills, which is one of the research questions in this paper [22].

Specifically of interest for the development of the complex skill of collaboration, Kim and MacDonough [23] report increased collaborative learning interaction through the implementation of video. The complex skill of presentation has been studied by De Grez et al. [24], finding learning through video to be more effective than extensive practice in the performance of presentations skills. The complex skill of information literacy has been studied by Frerejean et al. [25], finding the effective presentation of supportive information in the form of a video modeling example to cause a strong learning effect.

Concluding, video provides contextual and dynamic information, potentially leading to a richer mental model which can result in improved complex skill performance. However, in order to benefit from the qualities of video, it is important to specify design considerations that guide the synthesis of video modeling examples and a text-based analytic rubric.

In addition to the explored individual qualities of the text-based rubric and video, these media share qualities beneficial to the development of complex skills. Both the text-based rubric and video modeling examples have been shown to foster self-assessment, self-regulatory skills, and self-efficacy [1, 13, 16, 18, 26–31]. Self-regulated learning and self-efficacy may act as a motivational predictor of performance on complex tasks and its constituent processes, such as search, information processing and memory processes that affect learning [29, 32, 33]. Regulated learning is of particular importance for the performance of the complex skill information literacy as the learner constantly monitors and steers the information problem-solving process [34]. Self-regulated learning is also stressed by De Grez [33] as an important factor in the development of the complex skill of presenting. De Grez [33] found significantly increased learning gains as a result of implementing 'self-generated focused learning goals'. In addition to self-regulation, the self-efficacy fostered by both rubric and video is critical in the development of presentation skills and can result in significant learning gains [35].

Concluding, we expect that both rubric and video and have a positive effect on self-regulated learning, which contributes to complex skills mastery (Fig. 3).

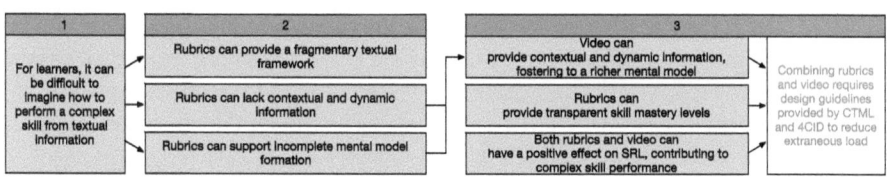

Fig. 3. The core summary of par. 1, 2 and 3

We now examine theory to facilitate the synthesis of video modeling examples and text-based rubric into a VER. The cognitive theory of multimedia learning (CTML) states that learners perform significantly better on problem-solving transfer tests through the combined use of video and text, as opposed to text alone [36]. Within this theory, several principles for the effective implementation of multimedia are described. The CTML principles aim to achieve three goals, namely to (1) reduce extraneous cognitive processing, (2) manage essential cognitive processing and (3) foster generative cognitive processing. First, reducing extraneous load is achieved by lightening the load of the instruction. This can be done by excluding extraneous information, signaling, highlighting and avoiding redundancy. Second, the goal of managing essential cognitive processing is to limit the intrinsic load of the multimedia instruction on the learners' cognitive capacity, preventing essential cognitive overload. Relying mainly on Paivio's [37] dual channel theory, this principle states that the visual channel is overloaded by simultaneously observing a video and reading on screen text. This can be remedied by offloading the visual processing of text to auditory processing by implementing narration. Third, to foster generative cognitive processing, the principles based on social cues are introduced. The personalization, voice and embodiment principles foster a social response, increasing active cognitive processing. According to dual channel theory, active cognitive processing is essential in transferring information from the sensory memory into working memory. This transfer fosters the quality of the learning outcome [38]. Social queuing suggests that active cognitive processing is fostered by the use of natural human voices, gestures and behavior in multimedia as opposed to artificial elements [36]. Another principle contributing to generative processing is embodied cognition [39]. Embodied cognition suggests that the physical engagement of performing a complex skill may foster learning by adding tactile, sensory information to the auditory and visual information provided in a multimedia instruction. Moreno's [40] Cognitive-Affective Theory of Learning with Media (CATLM) takes into account the auditory and visual senses described in CTML, while adding the tactile element of embodied cognition and the olfactory (smell) and gustatory (taste) senses. The CATLM proposes that educationally effective multimedia is not only a result of a cognitive perspective of CTML but also requires the self-regulated learning fostered by both rubric and video, motivation, affect and emotion to engage the learner into actively selecting and processing multimedia.

The stability of the multimedia effect has been recently studied, finding even a relatively simple multimedia implementation may be substantially and persistently more beneficial for retaining knowledge as compared to text alone [41]. However, the inherent complexity of a complex skill is challenging for the limited amount of working memory available to the learner. If the principles are implemented incorrectly, cognitive load is increased, and mental model creation is limited, impacting performance [22]. To ensure effective implementation of video and text, Mayer's [36] CTML relies upon dual channel theory, cognitive load theory and the integrated theory of text and picture recognition [36]. In addition to the principles of CTML, several studies have been done on partial aspects of instructional multimedia design. Eitel and Scheiter's [42] review consisting of 42 studies regarding text and picture sequencing states that it is helpful for comprehension if the medium that contains the least complexity is presented first. Presenting the least complex medium (either text or picture) first may facilitate processing of the complex information presented in the second medium. These findings are in line with both CTML and the 4CID model as the CTML's pre-training principle states that simple information can be used to prime the learner for complex learning [36]. The 4CID model ranges task classes and supportive information from simple to complex to accommodate the sequencing principle [43].

Concluding, theory presenting guidelines for synthesizing video and rubric into a design for the VER is rooted in CTML and 4CID. We expect these theories to allow us to manage the cognitive load of combining a rubric with video modeling examples and the inherent complexity of complex skills. CATLM can then be used to foster the learners' active regulation of the VER (Fig. 4).

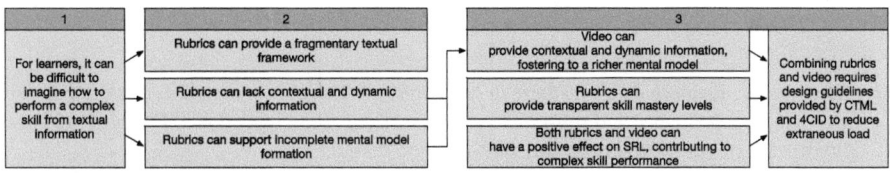

Fig. 4. The completed core summary

4 Conclusion

Analytic text-based rubrics mainly contribute to the development of complex skill on a cognitive level, providing rich feedback, anxiety reducing transparency and performance enhancing insight into the performance levels of a complex skill. However, despite these advantages, three problems regarding formative assessment of complex skills using rubrics where defined. The first problem indicates the aspect level of rubric assessment and states that a complex skill is more than the sum of its identified aspects. We expect that the implementation of video combined with text-based rubricswill help to unify the individual aspects of rubrics into a complex skill. We expect videoto address the first and third problem by 'filling the gaps' between the individual aspects of the complex skill as video provides the learner with the opportunity to personally encode semantic and dynamic information from the modeling example of the complex skill, supplementing the information provided by the rubric and providing personalization. The second problem indicates that the information that arubric provides when used as a systematic approach to problem solving by a learner is insufficient to form anaccurate mental model of a complex skill. For a video to convey the appropriate information to form a mental model, it is of importance that the modeling example conveys the mastery of the complex task in such a manner that the learner can encode the rich mental model found in the actions of the professional performing the task.

Concluding, we expect video to provide a rich enhancement to text-based rubrics for the specific use of (formatively) assessing complex skills. However, to ensure effective multimedia implementation it is of importance to adhere to CTML principles. In summary, we have taken a first step towards asynthesis of video and analytic text-based rubrics. We will focus further study on the development of design guidelines for effective implementation of video and analytic text-based rubrics for the (formative) assessment of complex skills in the Viewbrics project. Information on the Viewbrics project can be found on www.viewbrics.nl.

Acknowledgement. We would like to gratefully acknowledge the contribution of the Viewbrics project, that is funded by the practice-oriented research programme of the Netherlands Initiative for Education Research (NRO), part of The Netherlands Organisation for Scientific Research (NWO).

References

1. Panadero, E., Romero, M.: To rubric or not to rubric? The effects of self-assessment on self-regulation, performance and self-efficacy. Assess. Educ. Principles Policy Pract. **21**, 133–148 (2014). doi:10.1080/0969594X.2013.877872
2. Rusman, E., Martínez-Monés, A., Boon, J., et al.: Computer Assisted Assessment – Research into E-Assessment: Proceedings of International Conference, CAA 2014, Zeist, The Netherlands, June 30–July 1 2014. In: Kalz, M., Ras, E. (eds.), pp. 1–14. Springer International Publishing, Cham (2014)
3. Thijs, A., Fisser, P., van der Hoeven, M.: 21E Eeuwse Vaardigheden in Het Curriculum Van Het Funderend Onderwijs. Slo 128 (2014)
4. Janssen-Noordman, A.M., Van Merriënboer, J.J.G.: Innovatief Onderwijs Ontwerpen. Wolters-Noordhoff, Groningen (2002)
5. Van Merriënboer, J.J.G., Kester, L.: The four-component instructional design model: multimedia principles in environments for complex learning. In: The Cambridge Handbook of Multimedia Learning (2005). doi:10.1017/CBO9781139547369.007
6. Van Merriënboer, J.J.G., Kirschner, P.A.: Ten Steps to Complex Learning. Lawrence Erlbaum Associates Inc., New Jersey (2007)
7. Jonsson, A., Svingby, G.: The use of scoring rubrics: reliability, validity and educational consequences. Educ. Res. Rev. **2**, 130–144 (2007). doi:10.1016/j.edurev.2007.05.002
8. Westera, W.: Reframing contextual learning: anticipating the virtual extensions of context **14**, 201–212 (2011)
9. Matthews, W.J., Buratto, L.G., Lamberts, K.: Exploring the memory advantage for moving scenes. Vis. Cogn. **18**, 1393–1420 (2010). doi:10.1080/13506285.2010.492706
10. Panadero, E., Jonsson, A.: The use of scoring rubrics for formative assessment purposes revisited: a review. Educ. Res. Rev. **9**, 129–144 (2013). doi:10.1016/j.edurev.2013.01.002
11. Mertler, C.: Designing scoring rubrics for your classroom. Pract. Assess. Res. Eval. **7**, 1–10 (2001)
12. Brookhart, S.M., Chen, F.: The quality and effectiveness of descriptive rubrics. Educ. Rev. 1–26 (2014). doi:10.1080/00131911.2014.929565
13. Reynolds-Keefer, L.: Rubric-referenced assessment in teacher preparation: an opportunity to learn by using. Pract. Assess. Res. Eval. **15**, 1–9 (2010)
14. Andrade, H.G.: The effects of instructional rubrics on learning to write. Curr. Issues Educ. **4**, 1–39 (2001)
15. Schamber, J.F., Mahoney, S.L.: Assesing and improving the quality of group critical thinking exhibited in the final projects of collaborative learning groups. J. Gen. Educ. **55**, 103–137 (2006). doi:10.1353/jge.2006.0025
16. Andrade, H., Du, Y.: Student perspectives on rubric-referenced assessment. Pract. Assess. Res. Eval. **10**, 1–11 (2005). doi:10.1080/02602930801955986
17. Good, T.L.: Two decades of research on teacher expectations: findings and future directions. J. Teach. Educ. **38**, 32–47 (1987). doi:10.1177/002248718703800406
18. Panadero, E., Tapia, J.A., Huertas, J.A.: Rubrics and self-assessment scripts effects on self-regulation, learning and self-efficacy in secondary education. Learn. Individ. Differ. **22**, 806–813 (2012). doi:10.1016/j.lindif.2012.04.007
19. Wolters, C.A.: Regulation of motivation: evaluating an underemphasized aspect of self-regulated learning. Educ. Psychol. **38**, 189–205 (2003). doi:10.1207/S15326985EP3804_1
20. Kuhl, J.: A functional-design approach to motivation and self-regulation: the dynamics of personality systems and interactions. In: Handbook of Self-regulation, pp. 111–169 (2000)

21. Panadero, E., Romero, M., Strijbos, J.W.: The impact of a rubric and friendship on peer assessment: effects on construct validity, performance, and perceptions of fairness and comfort. Stud. Educ. Eval. **39**, 195–203 (2013). doi:10.1016/j.stueduc.2013.10.005
22. Gary, M.S., Wood, R.E.: Mental models, decision rules, and performance heterogeneity. Strateg. Manage. J. **32**, 569–594 (2011). doi:10.1002/smj.899
23. Kim, Y., McDonough, K.: Using pretask modelling to encourage collaborative learning opportunities. Lang. Teach. Res. **15**, 183–199 (2011). doi:10.1177/1362168810388711
24. De Grez, L., Valcke, M., Roozen, I.: The differential impact of observational learning and practice-based learning on the development of oral presentation skills in higher education. High. Educ. Res. Dev. **33**, 256–271 (2014). doi:10.1080/07294360.2013.832155
25. Frerejean, J., van Strien, J.L.H., Kirschner, P.A., Brand-Gruwel, S.: Completion strategy or emphasis manipulation? Task support for teaching information problem solving. Comput. Hum. Behav. **62**, 90–104 (2015). doi:10.1016/j.chb.2016.03.048. Manuscript Submission
26. Andrade, H., Buff, C., Terry, J., et al.: Assessment-driven improvements in middle school students' writing. Middle Sch. J. **40**, 4–12 (2009)
27. Brookhart, S.M., Chen, F.: The quality and effectiveness of descriptive rubrics. Educ. Rev. **1911**, 1–26 (2014). doi:10.1080/00131911.2014.929565
28. Efklides, A.: Interactions of metacognition with motivation and affect in self-regulated learning: the MASRL model. Educ. Psychol. **46**, 6–25 (2011). doi:10.1080/00461520.2011.538645
29. Schunk, D.H., Usher, E.L.: Assessing self-efficacy for self-regulated learning. In: Handbook of Self-Regulation of Learning and Performance, pp. 282–297 (2011)
30. Zimmerman, B.J., Kitsantas, A.: Acquiring writing revision and self-regulatory skill through observation and emulation. J. Educ. Psychol. **94**, 660–668 (2002). doi:10.1037/0022-0663.94.4.660
31. Van Dinther, M., Dochy, F., Segers, M.: Factors affecting students' self-efficacy in higher education. Educ. Res. Rev. **6**, 95–108 (2011). doi:10.1016/j.edurev.2010.10.003
32. Bandura, A.: Theoretical perspectives. In: Self-efficacy: The Exercise of Control, 604 pages. W.H. Freeman, New York (1997)
33. De Grez, L., Valcke, M., Roozen, I.: The impact of an innovative instructional intervention on the acquisition of oral presentation skills in higher education. Comput. Educ. **53**, 112–120 (2009). doi:10.1016/j.compedu.2009.01.005
34. Brand-Gruwel, S., Wopereis, I., Vermetten, Y.: Information problem solving by experts and novices: analysis of a complex cognitive skill. Comput. Hum. Behav. **21**, 487–508 (2005). doi:10.1016/j.chb.2004.10.005
35. De Grez, L., Valcke, M., Roozen, I.: The impact of goal orientation, self-reflection and personal characteristics on the acquisition of oral presentation skills. Eur. J. Psychol. Educ. **24**, 293–306 (2009). doi:10.1007/BF03174762
36. Mayer, R.E.: Multimedia Learning, 2nd edn. Cambridge University Press, New York (2009). doi:10.1007/s13398-014-0173-7.2
37. Paivio, A.: Mental Representations: A Dual Coding Approach (2008). doi:10.1093/acprof:oso/9780195066661.001.0001
38. Ayres, P.: State-of-the-art research into multimedia learning: a commentary on mayer's handbook of multimedia learning. Appl. Cogn. Psychol. **29**, 631–636 (2015). doi:10.1002/acp.3142
39. Skulmowski, A., Pradel, S., Kühnert, T., et al.: Embodied learning using a tangible user interface: the effects of haptic perception and selective pointing on a spatial learning task. Comput. Educ. **92–93**, 64–75 (2016). doi:10.1016/j.compedu.2015.10.011
40. Moreno, R., Mayer, R.: Interactive multimodal learning environments. Educ. Psychol. Rev. **19**, 309–326 (2007). doi:10.1007/s10648-007-9047-2

41. Schweppe, J., Eitel, A., Rummer, R.: The multimedia effect and its stability over time. Learn. Instr. **38**, 24–33 (2015). doi:10.1016/j.learninstruc.2015.03.001
42. Eitel, A., Scheiter, K.: Picture or text first? Explaining sequence effects when learning with pictures and text. Educ. Psychol. Rev. **27**, 153–180 (2014). doi:10.1007/s10648-014-9264-4
43. Van Merriënboer, J.J.G., Kester, L.: The four-component instructional design model: multimedia principles in environment for complex learning. In: Mayer, R.E. (ed.) The Cambridge Handbook of Multimedia Learning, pp. 104–148. Cambridge University Press, New York (2014)

Case Study Analysis on Collaborative Problem Solving Using a Tangible Interface

Dimitra Anastasiou[✉] and Eric Ras

IT for Innovative Services, Luxembourg Institute of Science and Technology,
5, avenue des Hauts Fourneaux, 4362 Esch-sur-Alzette, Luxembourg
{dimitra.anastasiou,eric.ras}@list.lu

Abstract. In our research within the "Gestures in Tangible User Interfaces" project we examine how gestural and speech behaviour while solving a collaborative problem on tangible interfaces has an effect on 21st Century skills. This paper presents an explorative case study where 15–22 year old pupils of a public school in Luxembourg solved a collaborative problem using a tangible user interface (TUI). The main goal of the study was to investigate the frequency and kinds of gestures and the link between the usage of gestures with collaboration and problem solving. At the end of the study, participants had to fill in three standard post-test questionnaires about workload and usability, the results of which are presented here and show that the use of a TUI as an educational medium was regarded as a straightforward and simple solution that many people could learn to use very quickly. This promising result is in line with our research vision, which is to include TUIs in a collaborative setting educational environment.

1 Introduction

Collaborative problem solving (ColPS) is a relevant competency which is extremely needed nowadays at workplace, but also in public and personal life. According to Tager (2013), the terms "collaborative problem solving", "cooperative work", and "group work" are used interchangeably in the education research literature to mean similar constructs. We follow the definition of Griffin et al. (2012), where ColPS refers to the abilities to recognize the points of view of other persons in a group; contribute knowledge, experience, and expertise in a constructive way; identify the need for contributions and how to manage them; recognize structure and procedure involved in resolving a problem; and as a member of the group, build and develop group knowledge and understanding.

In order to measure ColPS skills, our research project goal is to analyse the participants' gestural and speech behaviour while solving a ColPS scenario on a tabletop. Co-speech gestures are a very intuitive and expressive communication means and communication plays a crucial role in ColPS settings. Gestures have been often analysed in human-human conversational settings and also many gesture-based applications in human-computer interaction have been developed, but a systematic analysis of gestures on the application of TUIs is still lacking. Our aim is to incorporate gesture analysis as an assessment construct within the context of technology-based assessment (TBA).

The objectives of the explorative case study are to evaluate whether the developed e-assessment scenario is appropriate for the target group, keeping the pupils, on one hand, engaged and motivated, but on the other hand, with a quite high level of cognitive mental and temporal load. The results of the pilot study led to necessary improvements of the scenario and the assessment approach before conducting the main evaluation study with more participants.

The next section elaborates on the existing frameworks for ColPS and complex problem solving (CPS). Section 3 describes the design of the pilot study and Sect. 4 presents the data collection as well as the discussion of the analysed data with regards to workload, usability, and user acceptance of the TUI in an educational environment.

2 Related Frameworks

Noteworthy are the definitions not only of ColPS, but also CPS competences by the international educational Programme for International Student Assessment (PISA). The PISA 2015 CPS framework[1] identified three main *collaborative competences*:

1. Establishing and maintaining shared understanding;
2. Taking appropriate action to solve the problem;
3. Establishing and maintaining team organization.

The framework included also the so-called *CPS competencies*:

1. Exploring and understanding;
2. Representing and formulating;
3. Planning and executing;
4. Monitoring and reflecting.

Currently, in practice, PISA instantiates ColPS by having problem solving units including chat-based tasks, where students interact with one or more computer agents or simulated team members, to solve the problem at hand.

In our opinion, e-assessment of ColPS should rather involve a face-to-face setting with collocated people (see discussion in Petrou and Dimitracopoulou 2003; Dillenbourg 1999). Dillenbourg (1999) takes the unsatisfactory for him definition of *collaborative learning* as "a situation in which two or more people learn something together" and interprets in a different way the elements included in this definition. For example, whether "two or more" means a small group or a few hundred or "together" means face-to-face or computer-mediated. Our contribution towards collaborative learning is to conduct case studies where groups of three pupils solve a complex problem collaboratively on a new educational ICT medium, which is a tangible tabletop. One long-term objective of our project is to provide Interaction Design guidelines for ColPS using a TUI either for high-stake assessments (e.g. future PISA programmes) or in other settings where the aspect of learning by means of formative assessment is in focus. The well-established collaborative and CPS competencies

---

[1] https://www.oecd.org/pisa/pisaproducts/Draft%20PISA%202015%20Collaborative%20Problem%20Solving%20Framework%20.pdf, 06.06.16.

defined by PISA are still in place in our studies, but the distinct difference is that we offer a *collaborative* setting for ColPS rather an agent-based chat interaction, as in PISA.

Both collaboration and ICT technology are key elements in the Partnership for 21st Century Learning[2]. According to this framework (see Fig. 1), 21st century skills are different than 20th century skills primarily due to the emergence of very sophisticated ICT technologies. For example, the types of work done by people – as opposed to the kinds of labour done by machines – are continually shifting as computers and telecommunications expand their capabilities to accomplish human tasks.

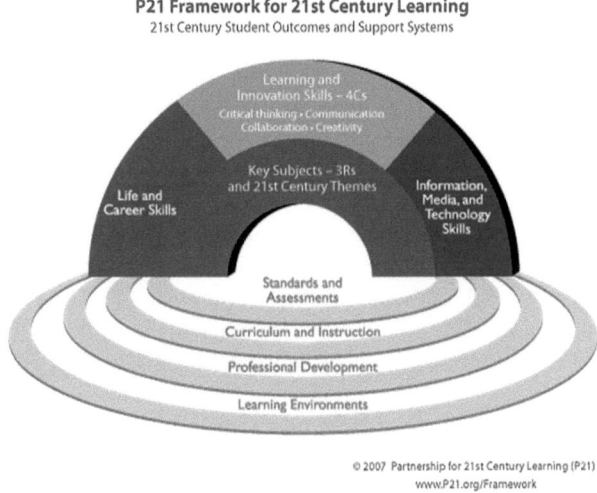

Fig. 1. 21st century student outcomes (arches of the rainbow) and support systems (pools at the bottom) according to the framework for 21st century learning

Binkley et al. (2012) compared a number of available assessment frameworks for 21st century skills which have been developed around the world and analysed them based on their learning outcomes in measureable form. Based on this analysis, they developed the so-called KSAVE model (*K*nowledge, *S*kills, and *A*ttitudes, *V*alues and *E*thics) with the following ten important skills (Table 1):

Csapó et al. (2012) reviewed the contribution of ICT technologies to the advancement of educational assessment. They described main tendencies for computer-based assessment in four continents (Asia, Australia, Europe, and USA) and with regards to challenge of assessment, they talked about five general situations, each of which poses different implications for the role that technology might play in assessment of twenty-first-century skills:

i. The first is characterized by domains in which practitioners interact with the new technology primarily using specialized tools, if they use technology tools at all;

---

[2] http://www.p21.org/, 06.06.16.

Table 1. The ten 21st century skills based on the KSAVE model

Ways of Thinking
1. Creativity and innovation
2. Critical thinking, problem solving, decision making
3. Learning to learn, metacognition
Ways of Working
4. Communication
5. Collaboration (teamwork)
Tools for Working
6. Information literacy (includes research on sources, evidence, biases, etc.)
7. ICT literacy
Living in the World
8. Citizenship – local and global
9. Life and career
10. Personal and social responsibility – including cultural awareness and competence

ii. The second situation is characterized by those domains in which, depending upon the preferences of the individual, technology may be used exclusively or not at all;
iii. The third situation is defined by those domains in which technology is so central that removing it would render it meaningless;
iv. The fourth situation relates to assessing whether someone is capable of achieving a higher level of performance with the appropriate use of general or domain-specific technology tools than would be possible without them;
v. The fifth situation relates to the use of technology to support collaboration and knowledge building.

In our user studies we focus on the fifth domain of technology according to Csapó et al. (2012) and the two skills of *Ways of Working* by Binkley et al. (2012) which are communication and collaboration.

3 Pilot User Study

In this section we refer to the general set-up of the study, particularly the demographic and social data of the participants (3.1), the micro-world scenario defining the task of the pupils (3.2), as well as the technical setup (3.3).

3.1 Participants

In March 2016 we ran a pilot case study with 15 participants in total split into 5 groups of three participants each. The students were between 15–22 years old (M = 17). Two groups were mixed gender (2 female & 1 male, 2 male & 1 female), two only female, and one only male. The language that participants talked to each other was Luxembourgish. The participants were coming sequentially and grouped as a team in the classroom where the study took place. They were advised to bring the consent forms which were given in advance, as both parents and/or legal representatives had to sign them too.

Apart from the consent form, the participants were not informed in advance in any way pertaining to the problem solving task they would deal with in this particular study.

3.2 Problem Solving Based on a Micro-World Scenario

In this subsection we refer both to the hardware and the software we used in order to design a collaborative problem solving scenario on a TUI. Our goal was that the selected scenario has a learning educational impact, but at the same time, is still enjoyable for the pupils of that age to keep them engaged in the scenario. The scenario does not require any previous knowledge from the pupils. This helps also to measure their learning effect independently after the user studies. Moreover, the overall task should not be difficult to solve, as it is not an IQ test or a typical exam *per se*; the goal of our project is to analyze their gestural and speech behavior while solving the ColPS on the tabletop.

As far as the hardware is concerned, the TUI employed for this pilot study is realised as a tangible tabletop (75 × 120 cm). Physical objects that have a specific reacTIVision[3] marker underneath can be recognized and manipulated on the table in order to explore different factors. reacTIVision is an open source, cross-platform computer vision framework for the tracking of fiducial markers attached onto physical objects, as well as for multi-touch finger tracking.

Our micro-world scenario was about building a whole power grid; this task is similar to tasks given in the PISA programme. The scenario is designed by means of a software framework TULIP (Tobias et al. 2015) which supports the instantiation of ColPS scenarios on TUIs via an XML file. In our pilot study, there was a picture from the Luxembourg city depicted on the TUI. The three pupils were provided with three physical objects (one object per participant) made out of cardboard, having a label A, B, and C on the top. These variables represented industrial facilities that produce electricity, i.e. a wind park, solar park, and a coal-fired power plant. The variables A, B, and C were used in order to prevent the potential previous knowledge of the pupils from solving the ColPS in a different way. The A, B, C objects could be placed, dragged, and rotated on the TUI. The factors that could be changed based on the placement and rotation of the objects[4] were: (i) power generation and (ii) CO_2 emission. The output values were data in gigawatt (gW) and in million tons (mT) of CO_2. In particular, when CO_2 increased, there was an animation on the TUI: the city got polluted and "foggier" due to the temperature rise created by the CO_2 emissions. Our micro-world scenario is based on several linear equations describing the relationship between the parameters. The scenario is shown in Fig. 2.

The instructions were orally given by an experimenter prior to the commencement of the study. After the instructions, the pupils had a practice session of three minutes to find out themselves how the objects affect the power generation and CO_2. After the practice session, they were asked to solve the following seven questions as presented in Table 2. The questions were visualized on the TUI itself and the experimenter manipulated a physical object to move from one question to the next one.

---

[3] http://reactivision.sourceforge.net/, 05.06.16.
[4] The dragging of the objects along the TUI did not have any effect.

Fig. 2. Collaborative problem solving on a tabletop

Table 2. Questions during ColPS in our user study

Q1	Which facility can produce the maximum power?
Q2	Which facility produces pollution (CO_2)?
Q3	Place one object on the table that produces the least power and no pollution.
Q4	Place as many objects as necessary on the TUI, so that power of 5,55 Gigawatt (GW) is produced in total.
Q5	Place as many objects as necessary on the TUI, so that power of 1,02 GW is produced in total.
Q6	Place as many objects as necessary on the TUI, so that pollution of 4 million tons (mT) pollution (CO_2) is produced in total.
Q7	Place as many objects as necessary on the TUI, so that power of 2,55 Gigawatt (GW) and maximum 2 mT CO_2 are produced in total.

3.3 Technical Set-Up

The required technical equipment was transported to the public school and two/three experimenters needed at least two hours to set it all up and calibrate the system. There were three cameras recording the study (see Fig. 3): (i) a Kinect 2.0[5] depth sense camera recording 3D data of the participants (ii) a standard video camera, (iii) a GoPro camera mounted on the ceiling. By means of Kinect 2.0 we developed an application that can automatically analyze object manipulation in real time with regards to *which object* has been manipulated *when, by whom* using *which hand*. A more detailed description of this application is outside the scope of this paper.

---

[5] http://www.xbox.com/en-US/xbox-one/accessories/kinect-for-xbox-one, 04.06.16.

Fig. 3. Technical set-up of the pilot user study in the classroom

4 Data Collection and Analysis

A multimodal corpus of video volume of about 2.5 h was collected for all five groups. The data collected through our user studies are of three types: (i) audio-visual recordings, (ii) 3D and logging data, and (iii) questionnaires. The audio-visual recordings are being annotated with the software ELAN (Wittenburg et al. 2006), a professional tool for the creation of complex annotations on video and audio resources. The Kinect 2.0 depth sense camera is used for recognition of the spatial position of the participants and their gestural interaction with the objects. Our software framework TULIP (Tobias et al. 2015) uses an abstraction layer to receive information from the computer vision framework and a widget-model based on model-control representations. In this paper we focus particularly on the third type of the collected data, the questionnaires. A general description of the questionnaires is presented in Subsect. 4.1 followed by the results of each questionnaire.

4.1 Questionnaires

After the end of the study, we handed three standard post-test questionnaires to the pupils, who filled them in at the classroom where the study took place. These three questionnaires were:

1. Nasa task load index (Hart 2006);
2. Attrakdiff (Hassenzahl 2004);
3. System Usability Scale (Brooke 1996).

The Nasa task load index is a subjective workload assessment tool with ratings on mental, physical, temporal demands, as well as own performance, effort and frustration. More precisely the questions included in the questionnaire are (Table 3):

Table 3. Nasa task load index questionnaire

Mental demand	How much mental and perceptual activity was required (e.g. thinking, deciding, calculating, remembering, looking, searching, etc.)?
Physical demand	How much physical activity was required (e.g. pushing, pulling, turning, controlling, activating, etc.)?
Temporal demand	How much time pressure did you feel due to the rate or pace at which the tasks or task elements occurred?
Performance	How successful do you think you were in accomplishing the goals of the task set by the experimenter (or yourself)?
Effort	How hard did you have to work (mentally and physically) to accomplish your level of performance?
Frustration level	How insecure, discouraged, irritated, stressed and annoyed versus secure, gratified, content, relaxed and complacent did you feel during the task?

The second questionnaire we used had the purpose to evaluate the usability and design of the micro-world scenario on the TUI. *Attrakdiff* is used to measure the attractiveness of a product or service in the format of semantic differentials. It consists of 28 seven-step items whose poles are opposite adjectives (e.g. "confusing - clear", "unusual - ordinary", "Good - bad"); see all items in Fig. 5. Each set of adjective items is ordered into a scale of intensity. Each of the middle values of an item group creates a scale value for pragmatic quality (PQ), hedonic quality (HQ) and attractiveness (ATT).

The third questionnaire is the System Usability Scale (SUS) questionnaire, a tool for measuring the usability of a product or a service. In our case "the product" was the tabletop employed in an educational environment. SUS consists of 10 items (see Table 4) with five response options (from strongly agree to strongly disagree).

Table 4. SUS questionnaire

Q1	I think that I would like to use this system frequently
Q2	I found the system unnecessarily complex
Q3	I thought the system was easy to use
Q4	I think that I would need the support of a technical person to be able to use this system
Q5	I found the various functions in this system were well integrated
Q6	I thought there was too much inconsistency in this system
Q7	I would imagine that most people would learn to use this system very quickly
Q8	I found the system very cumbersome to use
Q9	I felt very confident using the system
Q10	I needed to learn a lot of things before I could get going with this system

Results on Workload

The results (see Fig. 4) show that the workload on the performance was higher rated compared to the other categories. The performance was about "how successful do you think you were in accomplishing the goals". This result can be attributed to the fact that there was textual feedback ("You have reached the goal") about the task performance implemented in the ColPS scenario (see Table 2). The mental and temporal demand followed at the second and third place respectively. These results in combination with the results that the frustration level and the effort required were quite low show that the scenario selected for this age group was appropriate, keeping the pupils, on one hand, engaged and motivated, but on the other hand, with a certain level of cognitive mental and temporal load.

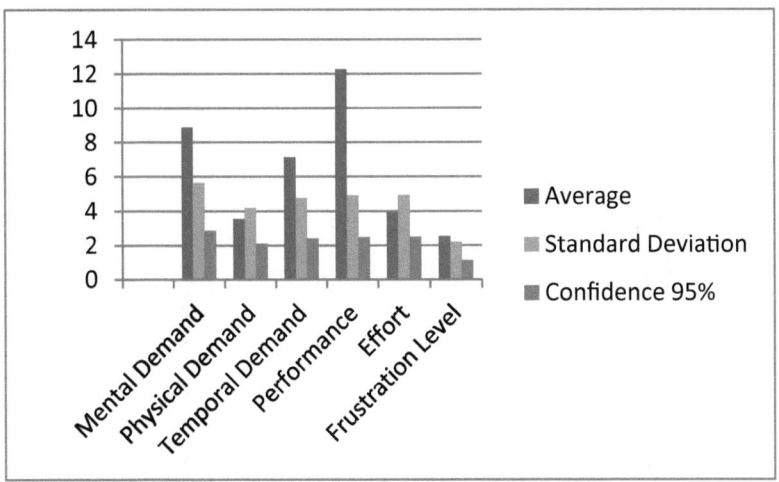

Fig. 4. Results of *the Nasa task load* questionnaire

Results on Attractiveness

The results of the Attrakdiff questionnaire are presented in Fig. 5. On the first picture (left), the various word pairs are presented along with their effect on pragmatic quality (PQ), hedonic quality (HQ) (including HQ-I and HQ-S) and attractiveness (ATT). The overall concept, solving a collaborative problem on the TUI, was ranked very *practical, creative,* and *good* by the participants. Striking is the result that although it was considered very *inventive*, it was also regarded as a *simple* and *straightforward* solution. This is a very positive result, showing that ICT technologies, and in our case the TUI, should not be intimidating, but rather employed to design straightforward scenarios where pupils can collaboratively solve a problem, get motivated, and even being provided with feedback. The portfolio presentation on the right side shows that the scenario balances between self-oriented and desired solution. This is the result of the scores of PQ (0.84) and HQ (1.30). The confidence rectangle shows that according to user consensus, the hedonic quality is greater than the pragmatic quality.

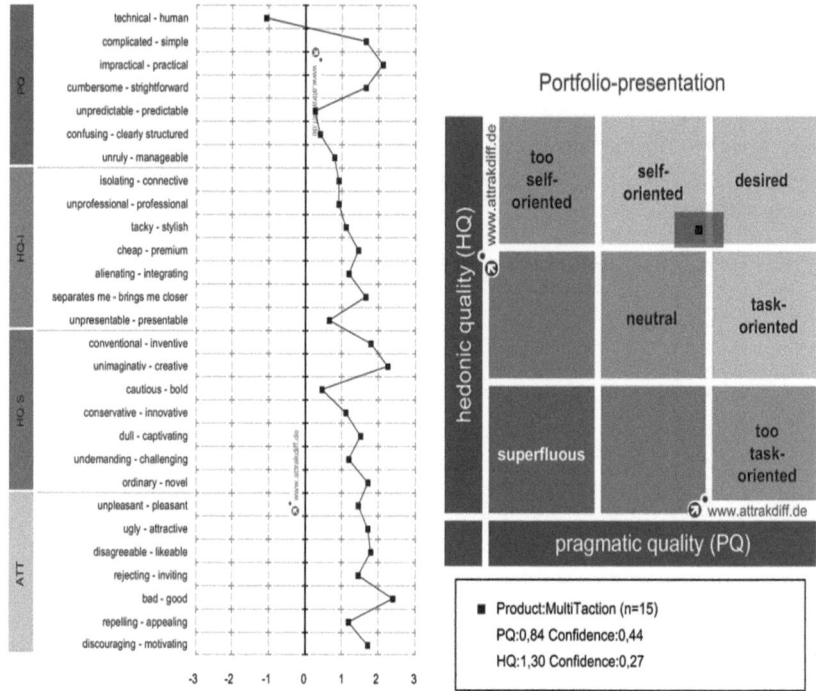

Fig. 5. Results of the *Attrakdiff* questionnaire

Results on Usability (SUS)

The resulting overall SUS score is 71.7, which is a very promising score. Figure 6 shows that the lowest SUS score was 45.0 given by the Participant 6 while the highest was 97.5

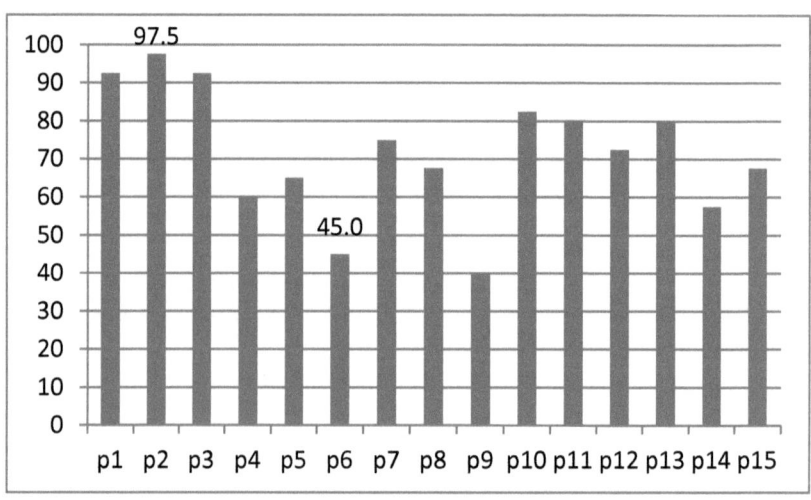

Fig. 6. Results of the *SUS* questionnaire

by Participant 2. Based on average scores, the question with the highest score was question 7 "I would imagine that most people would learn to use this system very quickly" which is a very optimistic and promising result particularly with regards to the applicability of TUIs in e-assessment in the future.

5 Conclusion and Future Prospects

An explorative case study with 15 participants was conducted in March 2016 in terms of the Marie Curie project GETUI. We spent about 2–3 months to design and implement an appropriate micro-world scenario on the TUI, which was employed in the study as a tangible tabletop, where physical objects can be used by the users to manipulate parameters in the scenario. In the design of the scenario, we took into account the PISA framework and its defined collaborative and complex problem solving competencies. Based on these requirements, our scenario was supposed to encourage and motivate users as well as to facilitate collaboration. These requirements were confirmed by the participants based on the results of the post-test questionnaires. Another significant result was that the micro-world scenario on the TUI was regarded as a simple and straightforward solution and in addition, that many people can learn to use the system very quickly.

As far as future prospects are concerned, we are currently annotating the video recordings. Our draft annotation scheme can be found at Anastasiou and Bergmann (2016). We distinguish between (i) physical or free hand movements (mainly pointing gestures) and (ii) manipulative gestures (dragging, rotating objects on the TUI). The former concern mainly human-human interaction, while the latter human-computer interaction. Speech transcription, conversational analysis as well as speech-gesture temporal alignment is planned at a later stage of the project. Moreover, a combination of our logging data from the TUI and the 3D data of the participants will help train our developed gesture recognition application which will decrease the time-consuming task of manual annotation in the long run.

Acknowledgment. This project has received funding from the European Union's Horizon 2020 research and innovation programme under grant agreement no. 654477. We thank the director as well as the pupils of the school Lycée Technique d'Esch-Sur-Alzette for the collaboration.

References

Anastasiou, D., Bergmann, K.: A gesture-speech corpus on a tangible interface. In: Proceedings of Multimodal Corpora Workshop, International Conference on Language Resources and Evaluation (2016)

Brooke, J.: SUS – a quick and dirty usability scale. In: Jordan, P.W., Thomas, B., Weerdmeester, B.A., McClelland, I.L. (eds.) Usability Evaluation in Industry. Taylor & Francis, London (1996)

Binkley, M., et al.: Defining twenty-first century skills. In: Griffin, P., McGaw, B., Care, E. (eds.) Assessment and Teaching of 21st Century Skills, pp. 17–66. Springer, Dordrecht (2012)

Csapó, B., et al.: Technological issues for computer-based assessment. In: Griffin, P., Care, E. (eds.) Assessment and Teaching of 21st Century Skills, pp. 143–230. Springer, Dordrecht (2012)

Dillenbourg, P.: What do you mean by collaborative learning. In: Collaborative-learning: Cognitive and Computational Approaches, vol. 1, pp. 1–15 (1999)

Griffin, P., Care, E., McGaw, B.: The changing role of education and schools. In: Griffin, P., McGaw, B., Care, E. (eds.) Assessment and Teaching 21st Century Skills, pp. 1–15. Springer, Heidelberg (2012)

Hassenzahl, M.: The interplay of beauty, goodness, and usability in interactive products. Hum.-Comput. Inter. **19**(4), 319–349 (2004)

Hart, S.G.: NASA-task load index (NASA-TLX); 20 years later. In: Proceedings of the Human Factors and Ergonomics Society 50th Annual Meeting, HFES, pp. 904–908 (2006)

Petrou, A., Dimitracopoulou, A.: Is synchronous computer mediated collaborative problem-solving 'justified' only when by distance? Teachers' points of views and interventions with co-located groups, during everyday class activities. In: Wasson, B., Ludvigsen, S., Hoppe, U. (eds.) Designing for Change in Networked Learning Environments, vol. 2, pp. 441–450. Springer, Dordrecht (2003)

Tager, M.: Computer-based assessment of collaborative problem-solving skills: human-to-agent versus human-to-human approach, Research report (2013)

Tobias, E., Maquil, V., Latour, T.: TULIP: a widget-based software framework for tangible tabletop interfaces. In: Proceedings of the 7th ACM SIGCHI Symposium on Engineering Interactive Computing Systems, pp. 216–221 (2015)

Wittenburg, P., et al.: ELAN: a professional framework for multimodality research. In: Proceedings of the 5th Conference on Language Resources and Evaluation, pp. 1556–1559 (2006)

Feedback Opportunities of Comparative Judgement: An Overview of Possible Features and Acceptance at Different User Levels

Roos Van Gasse[1(✉)], Anneleen Mortier[2], Maarten Goossens[1], Jan Vanhoof[1], Peter Van Petegem[1], Peter Vlerick[2], and Sven De Maeyer[1]

[1] University of Antwerp, Antwerp, Belgium
{roos.vangasse,maarten.goossens,jan.vanhoof,
peter.vanpetegem,sven.demaeyer}@uantwerpen.be
[2] Ghent University, Ghent, Belgium
{anneleen.mortier,peter.vlerick}@ugent.be

Abstract. Given the increasing criticism on common assessment practices (e.g. assessments using rubrics), the method of *Comparative Judgement* (CJ) in assessments is on the rise due to its opportunities for reliable and valid competence assessment. However, up to now the emphasis in digital tools making use of CJ has lied primarily on efficient algorithms for CJ rather than on providing valuable feedback. *Digital Platform for the Assessment of Competences* (D-PAC) investigates the opportunities and constraints of CJ-based feedback and aims to examine the potential of CJ-based feedback for learning. Reporting on design based research, this paper describes the features of D-PAC feedback available at different user levels: the user being assessed (assesse), the user assessing others (assessor) and the user who coordinates the assessment (*Performance Assessment Manager* (PAM)). Interviews conducted with different users in diverse organizations show that both the characteristics of D-PAC feedback and the acceptance at user level is promising for future use of D-PAC. Despite that further investigations are needed with regard to the contribution of D-PAC feedback for user learning, the characteristics and user acceptance of D-PAC feedback are promising to enlarge the summative scope of CJ to formative assessment and professionalization.

1 Introduction

In current diverse assessment practices, one assessor is responsible for grading assessees' performances one after another solely. This grading is generally done by means of rubrics[1], which are often lists of several criteria to grade competences. Then, the scores of the criteria are combined and a final mark is provided. Although this analytic method is widely spread among practitioners, the method has several issues

---

[1] In education, a rubric is a scoring aid to evaluate the quality of responses. Rubrics Rubrics usually contain evaluative criteria, quality definitions for those criteria at particular levels of achievement, and a scoring strategy (Popham 1997).

(Pollitt 2012). First, rubrics divide the competence in several artificial dimensions, assuming strict boundaries between all dimensions. However, competences cannot be divided into several dimensions, since these dimensions can overlap. Moreover, the dimensions combined cannot contain the entire competence (Sadler 2009). Second, non-competence related aspects can influence the judgment, such as mood and time of the day (Bloxham 2009). Even when assessors are trained, such issues still affect the assessment (Crisp 2007). Last, even though assessees are assumed to be assessed independently, they are still compared with each other during grading. Therefore, absolute judgment is an illusion (Laming 2004).

These biases have recently resulted in an increased popularity of an alternative assessment practice; Comparative Judgment (CJ). In this method performance is assessed by comparing representations[2] of the same competence (Pollitt 2004), since comparisons are easier than grading (Thurstone 1927). The intent is that, in contrast to rubrics, various assessors compare representations and decide which of them is better with regard to a certain competence. By means of statistical modeling, multiple comparisons of multiple assessors result in a rank order in which representations are scaled relatively from worst to best performance (Bramley 2007). Major strengths of this method are the potential to result in a high level of reliability since CJ depends on direct comparisons (Pollitt 2012), ruling out assessors' personal standards, limiting non-task relevant influences, and feeling more natural for assessors compared to other methods (Laming 2004).

Up to now, a small number of digital tools has been developed in which CJ is the central assessment method (e.g. NoMoreMarking, e-scape). However, the majority of these tools are focused on assessing rather than learning. The emphasis in these tools lies on the (refinement of the) statistical algorithm behind CJ in order to obtain a reliable rank order more efficiently. Currently, there is lack of digital tools using CJ as a method that serves users of the tool with valuable learning possibilities. Therefore, the project developing a *Digital Platform for the Assessment of Competences* (D-PAC) investigates the opportunities and constraints of CJ-based feedback and aims to examine the potential of CJ-based feedback for learning. The D-PAC platform provides feedback at three levels: assessee level, assessor level and *Performance Assessment Manager* (PAM) level. Assessee level is defined as the level of assessed users, for example students in schools. Here, the main goal of feedback is to enhance students' performance. Assessor level is defined as the level of assessing users, for example teachers in schools. At this level, the aim of feedback is to inform assessors of their performance in assessing. The PAM level is the level on which the whole assessment is coordinated, for example the head teacher or school board. The coordination role of the PAM implies that PAMs are also in charge to provide feedback towards assessees and assessors. D-PAC is built in a way that it is possible to adjust every little detail of feedback. Thus, PAMs can decide feedback features that are necessary for assessees and assessors and adjust the feedback reports towards both groups.

---

[2] A representation is a medium that reflects a certain competence, for example an essay for writing skills, a video for acting skills, a drawing for artistic skills,....

In this paper, we aim to present the possible feedback features of D-PAC at assessee, assessor and PAM level. Since the feedback research in D-PAC is ongoing, we will discuss some preliminary findings about user experiences for each feedback feature. The following research questions will guide our investigation:

1. Which feedback features are possible using CJ assessment?
2. How do feedback features of D-PAC contribute to users' acceptance of feedback at all levels?

2 Theoretical Framework

Feedback is considered as a powerful instrument to enhance performances (Hattie and Timperley 2007). In feedback literature can be distinguished between feedback characteristics (i.e. its focus, content and presentation) and feedback acceptance (i.e. its relevance, validity and reliability and accessibility and user friendliness) (Hattie and Timperley 2007). Both are important to initiate learning opportunities.

2.1 Feedback Characteristics

According to Hattie and Timperley (2007), feedback can be focused at different layers of performance. At the first layer ("self"), feedback is provided to the user as a person and is not necessarily linked to the performance itself or task goals for future learning. Next, indications can be provided for the evaluation of the task (e.g. what went good and what went wrong). Furthermore, feedback on the process focuses on which steps were followed to complete the task. Lastly, feedback on the regulation is related to processes within the feedback user (e.g. self-assessment) (Hattie and Timperley 2007).

Regardless of the focus of feedback, effective feedback should answer three questions: (1) where am I going? (2) how am I going? and (3) where to next? The first question ("where am I going?") relates to the task or performance goals. The second question ("how am I going?") should provide information on one's relative standing towards a certain standard. The last question ("where to next?") provides opportunities for a greater chance of learning. Additionally, feedback should reflect a certain amount of time invested in the process (van der Hulst et al. 2014). Differences in feedback content can occur depending on the source of feedback (e.g. peers or teachers) (Nicol and MacFarlane-Dick 2006).

A last feedback characteristic of great importance is the presentation of feedback. Feedback should be readable, neat, comprehensible, individualized and provided within an appropriate time span to users (Nicol 2009; Shute 2008; Xu 2010). Digital feedback has the potential to present feedback in a useful way for users and as such to boost its quality (van den Berg et al. 2014).

2.2 Feedback Acceptance

Feedback can be a powerful tool to initiate learning at different levels in organizations as well (Verhaeghe et al. 2010). However, the contribution of feedback to users' learning can only be reached if feedback is accepted (Anseel and Lievens 2009; Rossi and Freeman 2004). A prerequisite of feedback acceptance is that feedback is perceived as useful and fair (Schildkamp and Teddlie 2008; Visscher 2002). This indicates that it is crucial to concern certain characteristics in feedback design to enhance the probability of feedback acceptance and thus future feedback use and user learning (see Fig. 1). In what is next, we will describe the need of feedback to be relevant, reliable and accessible for users.

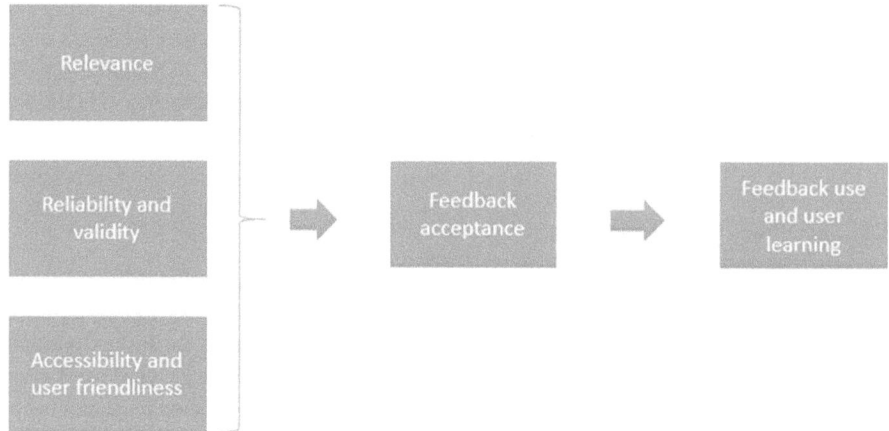

Fig. 1. Model of feedback acceptance

Research has indicated that the perceived relevance of feedback is important for the way people will use it (Verhaeghe et al. 2010a). The importance of feedback relevance is reflected in the fact that people do not tend to use data which is not perceived as necessary in serving individual or organizational purposes (Vanhoof and Mahieu 2011; Vanhoof et al. 2009).

Furthermore, the actual or perceived reliability and validity of the information concerned also influences whether or not feedback is used at different user levels (Kowalski and Lasley 2009; Pierce and Chick 2011; Schildkamp and Teddlie 2008). A certain type of information is only used if users have the feeling that this information is correct, covers the right things and represents them accurately.

Lastly, literature has indicated that it is crucial that the feedback system is built in a user friendly and accessible way so that information becomes easily available (Schildkamp and Teddlie 2008). Additionally, feedback should be ready for use (i.e. users should not have to further manipulate data). Good feedback systems are built in a way that reduce extra work load and enhance efficiency (Young 2006).

3 Method

In order to investigate our present research questions, we started from a design research lens. This means that implementation in organizations was crucial for the (further) development of D-PAC feedback (Collins et al. 2004; Hevner 2007).

In a first phase, the team[3] investigated which feedback features were possible with regard to CJ assessments. Then, with the team, several brainstorm sessions were held to decide on interesting features to include in D-PAC feedback. The time frame was set up for when these features needed to be implemented.

Next, the team discussed the features mentioned in the brainstorm sessions with a group of nine practitioners (education and HR context). These discussions were used to decide which features would be included in D-PAC feedback.

Subsequently, a draft feedback report was developed for the different feedback levels (assessee, assessor, PAM). The draft reports were discussed within the team and again presented to a group of practitioners. Remarks mentioned by the group of practitioners were taken into account for the further development of the feedback report.

Then, several assessments via D-PAC were set up in various organizations in education and HR contexts (try-outs). Together with the organization in which the assessment was unrolled, decisions on which feedback would be provided were made for the three levels (assessee, assessor, PAM). After feedback was delivered, feedback users (see

Table 1. Overview of try-outs in D-PAC in which feedback was provided.

	Competence	Domain	Assessees	Assessors	PAM	Interviewed
1	Argumentative writing	Education	136 high school students	68 teachers and teachers in training	10 head teachers	36 assessees and 6 PAMs
2	Writing formal letters	Education	12 bachelor students	11 teachers in training	1 lecturer	11 assessors
3	Visual representation	Education	11 bachelor students	13 teachers	1 lecturer	1 PAM
4	Analysing skills	Education	84 master students	4 professors	1 professor	1 PAM
5	Moodboards	Education	60 bachelor students	60 bachelor students 4 lecturers	1 lecturer	60 assessees 4 assessors 1 PAM
6	Self-reflective skills	Education	22 master students	9 lecturers	1 professor	9 assessors 1 PAM
7	Leadership skills	Selection	20 candidate school leaders	6 professionals (jury)	1 chair jury	6 assessors 1 PAM

---

[3] D-PAC is funded by the Flemish Agency of Innovation and Entrepreneurship and is a partnership between researchers of three Flemish universities (Universiteit Antwerpen, Universiteit Gent and Vrije Universiteit Brussel (iMinds)).

Table 1) were questioned (interview or focus group) on their understanding of the different features in the feedback report and their use of the feedback report. The individual and focus group interviews were transcribed ad verbatim. Next, summaries of the transcriptions were synthesized, compared and discussed by several members of the D-PAC team. Subsequently, conclusions were drawn about the acceptance of feedback at different user levels and decisions were made on the feedback features in future try-outs.

Thus, the try-outs were used to increase our understanding of how D-PAC feedback was accepted by its users. In the upcoming years, our findings of the try-outs will be used to further develop D-PAC feedback. Table 1 provides an overview of the try-outs in which feedback was provided. Different competences were assessed (e.g. writing, moodboards, …) in education and (education) selection domain. Assessees varied from high school students to candidate principals. Assessors and PAMs were mostly teachers (in training) or lecturers.

4 Results

Table 2 provides information on the features that are available in D-PAC feedback Since PAMs manage the assessments, they can also decide which features are available for certain roles (assessee, assessor or PAM). For example, a PAM can decide to

Table 2. Feedback features per level.

Feedback feature	Assessee	Assessor	PAM
General statistics			
Reliability	Y	Y	X
Mean comparisons per assessor	Y	Y	X
Mean comparisons per representation	Y	Y	X
Total time spent on assessment (overall)	Y	Y	X
Mean time per comparison (overall)	Y	Y	X
Total time spent on assessment (personal)	Y	Y	X
Misfit statistics			
Misfit assessor (personal)		Y	X
Misfit assessor (overall)			X
Misfit representations (overall)		Y	X
Rank order	Y	Y	X
Specific feedback	Y One assesse/all assessees	Y One assesse/all assessees	Overall

X = standard
Y = PAM can decide to switch the feature on or off

(not) insert a rank order in the feedback to assessees. Another aspect of PAM feedback is that this type of feedback is available during the assessment as well as at the end of the assessment. This way, PAMs can monitor aspects of the assessment in order to decide about the progress of the assessment (e.g. reliability). In this section, we will describe the different features of D-PAC feedback, following the structure of Table 2.

4.1 General Statistics

Prior to the feedback features described below, users are provided with general information regarding the (finished) assessment. Therefore, an overview of the assessment is given in which the number of assessors, comparisons and representations of the assessment is included.

Reliability of the Assessment. The advantage of CJ compared to rubric assessing is that the work of multiple assessors can be evaluated in terms of reliability (Pollitt 2012). Therefore, the first feature of D-PAC feedback is a reliability measure for the assessment. Because the Rasch model is used to analyse the CJ data, the Rasch separation reliability or Scale Separation Reliability (SSR; Bramley 2015) can be calculated. The measure represents the amount of spread in the results that is not due to measurement error (McMahon and Jones 2014). According to Anshel et al. (2013) the SSR is an indication for how separable the representations are on the final scale of the assessment. So, a small measurement error implies that the relative position of the items on the scale is quite fixed (Andrich 1982)

Like a Cronbach's Alpha value, the SSR is a value between 0 and 1. The higher the value, the higher the chance that the same rank order would be reached in a new assessment with the same set of assessors. Thresholds of SSR are similar to those of a Cronbach's Alpha value (i.e. 0.70 for a reasonable reliability and 0.80 for a good reliability).

Up to now, we have limited information about the value dedicated to the reliability in general. Current experiences with assessors and PAMs indicate that the reliability measure is evaluated in both groups. However, in these experiences reliability measures were each time sufficient (i.e. at least 0.70) so that no information is gained on implications of a failing reliability for how the assessment (and CJ) is evaluated in both groups. Thus, we do not (yet) have insight into how the reliability measure affects users' acceptance of feedback if the reliability is not sufficient.

First experiences with the reporting the reliability measure in assessee feedback showed mixed results. Some assessees found it interesting to know the reliability of the assessment, others did not care about it and were only interested in their own performance and the specific feedback they received.

Number of Comparisons per Assessor and per Representation. In order to obtain a reliable assessment, choices need to be made about the number of comparisons that will be carried out by each assessor. Therefore, D-PAC feedback provides information on the number of comparisons in the assessment.

At assessee level, the number of comparisons of assessees' representation and the number of assessors who judged their representation can be included in feedback. Although some assessees perceived these measures as informative, others perceived this type of feedback as unnecessary.

At assessor level, the total number of comparisons, the mean number of comparisons per assessor, and one's own number of comparisons is given. Until now, it remains unclear whether users find this information relevant and whether or not this feature contributes to the acceptance of feedback among assessors.

For PAMs, two types of information are available on the number of comparisons. First, information on the number of comparisons is generated at assessor level. This means that the PAM is provided with insights into the number of comparisons made per assessor. Together with information on the time investment of each assessor, this type of information can help PAMs to evaluate assessors' workload. Second, information on the number of comparisons is generated at representation level. This means that PAMs are provided with information on how many times a representation is compared to another one. The number of comparisons per representation can influence the reliability of the assessment. Therefore, this statistic can be taken into account when the reliability is low. Up to now, little information is available on how the number of comparisons is used by PAMs. However, after one particular try-out, the PAM asked us to examine whether the reliability kept on increasing after a certain stage of the assessment. This information was important for the PAM in order to evaluate the assessment in terms of efficiency in order to improve the reliability costs balance in future use of CJ assessments.

Time Investment. D-PAC provides indications of the time invested in the assessment. This comprises calculations of the duration of all comparisons. Several descriptive statistics on the time investment are generated, such as minimum and maximum duration of a comparison, and median and mean duration across comparisons.

At the level of the assessor, D-PAC provides an overview is provided of the invested time of an assessor within and across comparisons. These statistics are indicative for one's efficiency by comparing the mean time per assessor spend per comparison with one's own mean time spend per comparison. Also the total time spend on the assessment can be informative as the subjective idea of the total time is not always accurate. Up to now, assessors have pointed some of the time statistics as informative but not necessary.

At PAM level, an overview is generated of the total time investment in the assessment within and across assessors. The time investment statistic was included in PAM feedback so that PAMs have insight into the efficiency of the assessment. The statistic can have the potential to serve PAMs with information on *inter alia* what type of assignments are suitable for CJ, less efficient assessors in CJ or the efficiency of a CJ assessment compared to other assessment methods used. Up to now, PAMs have indicated that the time investment statistic is interesting when running an assessment, but have not yet provided us with insights into how the statistic would be used during and after an assessment.

4.2 Misfit Statistics

Comparative judgement assessments provide the opportunity to include misfit statistics in feedback. Misfit statistics are based on Rasch modelling (1960; 1980), which uses chi-squared goodness of fit statistics to quantify how far judgements are different to what the model predicts. Aggregating these estimates can produce a measure of how much judges deviate from the group consensus or how ambiguous a representation is for the assessor group. There are two types of misfit statistics: infit and outfit. In D-PAC the infit was included given it being less prone to occasional mistakes (Linacre and Wright 1994). In short, the misfit statistic can be described as a quantification of an unexpected choice of an assessor in a specific comparison given the model expectation (Lesterhuis et al. 2016). A misfit is identified if the infit estimate lies minimum two standard deviations above the mean (Pollitt 2012) (see Fig. 2).

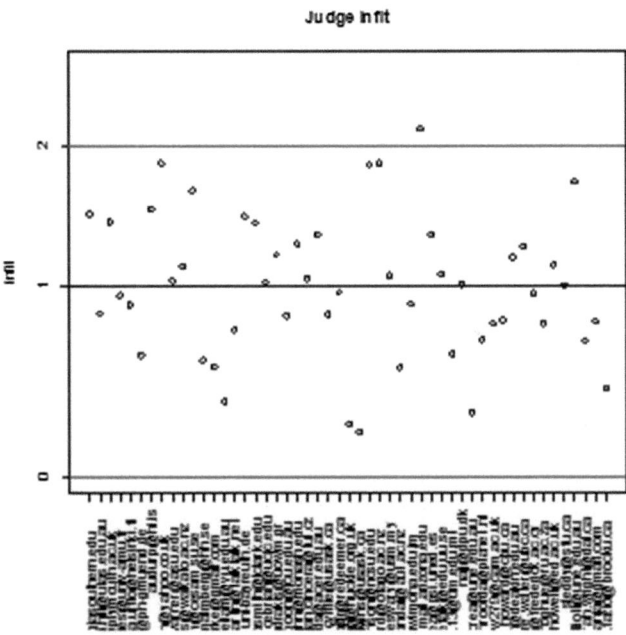

Fig. 2. Judge infit.

A misfitting assessor implies that this assessor generally makes judgements that differ from the consensus. This could be explained by differences in conceptualization (Bramley 2007) or finding certain aspects of a competence more important than other assessors (Lesterhuis et al. 2016). Excluding misfitting assessors can lead to an increased reliability of the rank order (Pollitt 2004).

A representation misfit implies that assessors make inconsistent judgements regarding this representation (Pollitt 2004). Unusual content of representations regarding the performance can lead to representations being more difficult to judge (Bramley 2007; Pollitt 2004).

Misfit statistics are included in D-PAC feedback for assessors and PAMs, and can initiate valuable processes of professionalization. When a PAM notices a misfitting assessor, discussions between assessors on how certain competences are conceptualized can be organized in order to align this conceptualization among them. The same rationale goes for misfitting representations, which can be discussed in order to reach a consensus about these representations. This again can contribute to the assessment practice among assessors.

We do find that assessors and PAMs perceive misfit statistics as an interesting feature of D-PAC feedback. Assessors have indicated that they are often looking for data which tell them 'how they are doing' or 'if they judge in line with colleagues' during assessments. PAMs value misfit statistics because they provide them with a quality check.

Up to now, we have gained limited information about the use of misfit statistics by means of the try-outs. In a single try-out, a PAM asked to exclude a misfitting assessor in order to increase the reliability of the assessment. Currently, a try-out is ongoing in which misfit statistics are used for professionalization of assessors.

4.3 Rank Order

In CJ assessments, a rank order is generated that represents the quality of representations in the light of the assessed competence (Bramley 2007). The rank order is derived from the multiple comparisons assessors made. Contrary to common assessment methods, the rank order is relative instead of absolute. Based on these, the Bradley-Terry-Luce model (Bradley and Terry 1952; Luce 1959) transforms the comparisons into the odds of winning a comparison with a random other representation (Pollitt 2012). Next, these odds are translated into a rank order. Confidence intervals can be represented on the figure, to indicate the 95% confidence intervals of every performance. The rank order can be included at all feedback levels (Fig. 3).

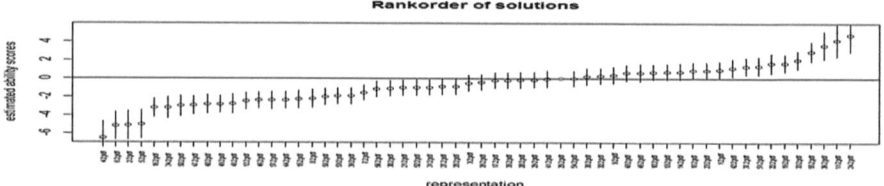

Fig. 3. Rank order.

For assessees, providing the rank order gave mixed results. Some assessees perceived this measure as informative, since they could compare their own results with their peers' result. However, others stated that this is very relative, and it did not give an indication of their score or mark.

For assessors, the rank order is informative since it obtains the direct result of their work during the assessment. Assessors who already used D-PAC have generally indicated that comparing representations makes them curious about the rank order of the representations. They like to check whether the representations they assessed as good also get a good ranking. However, assessors have also mentioned that they would like to get immediate results of their judgements, for example by receiving their personal rank order.

For PAMs, the rank order is the central outcome of a CJ assessment. Depending on the assessment purposes (formative or summative), how PAMs deal with the rank order can differ. Up to now, rank orders generated by D-PAC have been used in various ways by PAMs. For formative purposes, the rank order for example has served as a guide to illustrate qualitative and less qualitative assignments. For example, in the moodboards try-out, a lecturer in higher education discussed aspects of moodboards that were higher and lower positioned on the rank order. For summative purposes, PAMs have for example used the rank order to score assignments or to select candidates in a selection procedure. For example, in this selection procedure, the first 14 candidates were selected for a training for future principles. In the try-out on debriefing notes, the first and the last representation on the rank order was discussed and scored among the assessors. Subsequently, they gradually graded each representation following the rank order.

4.4 Specific Feedback

In D-PAC, there are two ways to provide specific feedback for assessees. First, it is possible that PAMs ask assessors to provide general feedback for each representation in a comparison (e.g. *"Task A is good on this competence, however it still needs more ..."*). Second, PAMs can decide to include a feedback frame for assessors in which positive and negative aspects of each representation within a comparison are explicated (e.g. *"Task A is good because...; Task A is not good because..."*).

The aim of both feedback features is to provide assessees with insights into the strengths and weaknesses of their task. Because the specific feedback demands a greater time investment for assessors during the assessment, we investigated the necessity for specific feedback for assessees.

Our results indicate that D-PAC feedback is perceived more relevant, reliable and fair when specific feedback was included, independent of the type of feedback (i.e. receiving general feedback on the task or specified positive and negative aspects). Therefore, the specific feedback towards assessees strongly contributes to assessees' acceptance of D-PAC feedback. Additionally, assessees believe that specific feedback contributed to their learning[4]. Compared to the classic method of providing feedback (e.g. via comments in Word), specific feedback out of D-PAC is preferred when it comes to improving competences for future assignments. Providing both positive and negative feedback via D-PAC has been perceived as valuable, since in the classic

---

[4] For more information, see Mortier et al. 2016.

feedback method (e.g. via comments in word), assessors are not inclined to insert positive comments. Content analysis confirmed assessees' perceptions, showing a significantly higher amount of motivational comments and learner-oriented feedback via D-PAC.

5 Discussion and Conclusion

Comparative Judgement (CJ) has the potential to serve feedback opportunities that provide learning chances for different types of users. Nevertheless, digital tools for CJ assessment focus primarily on how to create a reliable rank order more efficiently and pay limited attention to the creation of CJ based feedback. The project *Developing a digital Platform for the Assessment of Competences* (D-PAC) addresses this shortcoming by investigating CJ based feedback opportunities. In the current paper, an overview has been given of feedback opportunities for users being judged (assessees), users who judge (assessors) and users who coordinate the assessment (Performance Assessment Managers: PAMs). Additionally, D-PAC users' acceptance of feedback has been described.

Considering how D-PAC feedback is constructed at different levels (i.e. assesse, assessor and PAM), the feedback can provide users at all levels with features which have the potential to affect user learning. Firstly, several embedded feedback characteristics may create valuable learning opportunities for users. At all user levels (i.e. assesse, assessor and PAM), feedback is focused at the task- and process level, which are important for user learning (Hattie and Timperley 2007). At assessee level, the rank order can be used to gain insight in the quality of performance for a certain task compared to peers. Furthermore, assessees can be provided with insights of what was good and what needs to be improved with regard to the task (i.e. specific feedback). Strategies can be proposed on how to improve the current task or to complete a new, related task in a better way. At assessor and PAM level, misfit statistics provide indications on the quality of the judgements made in the assessment (i.e. the task of assessors).

Secondly, content-wise all ingredients are available in D-PAC to provide users with answers on (1) "where am I going?, (2) "how am I going?" and (3) "where to next?" (Hattie and Timperley 2007). Performance goals, process evaluation and improvement actions can be the result of a self-regulating process by means of D-PAC feedback. Assessees can for example use the rank order to look at good examples of peers to set objectives for future tasks and to improve their performances. At the level of assessors, the same rank order can be used to think about how the competences of assessees can be improved (in case the assessor is the actual tutor or lecturer of assessees) or one's misfit score can be a means of reflection (e.g. "What makes my judgements different to those of other assessors and what can I do to contribute to a better alignment in the judgement process?"). At PAM level, a trade-off between time investment and reliability can be made to set goals for future assessment in terms of assessment efficiency or misfit statistics can be used to introduce professionalization trajectories for assessors.

Lastly, several presentation issues are overcome. At assessee level, issues with readability (e.g. difficult handwriting) are overcome and feedback is individualized, which would increase assessees' engagement towards feedback (Xu 2010). The overall feedback that needs to be provided in D-PAC results in higher-order feedback (Bouwer et al. 2016). Also at assessor level, feedback is individualized (e.g. one's own misfit score is available) in a comprehensible way. At all levels, feedback is provided within a reasonable time span. Although assessees need to wait for feedback until the assessment has ended, assessors are provided with feedback timely after they finished their comparisons and PAMs are able to request feedback during the assessment as well. A footnote to be made in this regard is that the PAM coordinates feedback for assessees and assessors. Thus, timely feedback depends on whether or not the PAM decides to provide it timely.

In short, feedback characteristics that are found to be crucial for user learning are embedded in D-PAC feedback.

Next to feedback characteristics, users' acceptance of feedback has implications for the feedback's potential for learning. General acceptance of D-PAC feedback is reached among the different users (assessees, assessors, PAMs), which is a first crucial step for future use of D-PAC feedback (Anseel and Lievens 2009; Verhaeghe et al. 2010; Schildkamp and Teddlie 2008; Pierce and Chick 2011). At all user levels, D-PAC feedback incorporates features that are perceived as relevant, reliable and valid. Moreover, the features are easily accessible and perceived as user friendly. At the level of assessees, in particular the specific feedback is perceived as valuable. The way feedback needs to be formulated in D-PAC (holistically and higher order) provides assessees with insights in how they need to develop their competences in order to better accomplish future assignments. At the level of the assessor and the PAM, the positive reactions on the misfit statistics are promising. For assessors, in most assessments, information on how they accomplish their assessment practice is rare. In D-PAC, misfit statistics provide assessors with data to evaluate 'how they are doing'. At PAM level, misfit statistics inherit an assessor quality check for PAMs. When a misfitting assessor is identified by a PAM, attempts can be made to improve the assessing process and introduce discussions regarding how the judging task is approached among assessors.

Up to now, the D-PAC tool has developed promising feedback features for assessees, assessors and PAMs. However, limitations still remain in the opportunities for examining the impact of the different feedback features. Given that until now, try-outs were single occasions to use the D-PAC tool, feedback was given after the assessment at all three levels. Each time, the feedback was discussed afterwards and feedback acceptance was analyzed, but no follow up of feedback with a view to future assessments was possible in the organizations. Therefore, we were able to gain insight into the user satisfaction of feedback, but not in what users actually do with feedback and in how feedback drives learning. Thus, knowledge on how useful CJ feedback in general and D-PAC feedback in particular can be for users in individual and organizational learning processes is still lacking.

The current lack of knowledge on how D-PAC feedback is used and the contribution of D-PAC feedback to individual and organizational learning makes this a research area that should be addressed in future research. At the level of assessees, more research is needed in how the current feedback features, and in particular specific feedback, can initiate learning. It is necessary to gain insight into if and how this specific feedback is valuable for assessees in order to improve their competence under assessment. At the level of assessors, questions remain regarding the use of different feedback features, in particular misfit statistics, and its learning results. An interesting option would be to examine how assessors use misfit statistics and how the use of this data contributes to their personal development regarding the assessing task. At PAM level, knowledge on how PAMs use D-PAC feedback at different stages of the assessment would be useful. Therefore, it is needed that PAMs can be followed in assessments that are unrolled in D-PAC. This would provide information on how PAMs monitor and use the different feedback features. Also, future research should make attempts to address the contribution to organizational learning that PAM feedback can have. It would be valuable to obtain knowledge on how the different features of feedback can improve assessment strategies in organizations. Thus, several options remain in the further exploration of CJ based feedback in general and D-PAC feedback in particular.

The interesting options for feedback at assessee, assessor and PAM level reframe the use of CJ for assessments in schools and organizations. Whereas CJ has been seen as a valuable method to obtain a reliable method in an efficient way (Pollitt 2012), insights into the potential feedback features can broaden this lens. Instead of a useful method for summative assessment, in which one needs to select the best candidates in a reliable way, D-PAC feedback can also be seen as convenient for formative assessment. The CJ method has a lot of potential with regard to feedback features that potentially initiate learning at different user levels in organizations. Despite the limited research opportunities of feedback via D-PAC so far, promising feedback features are available in D-PAC to enlarge the summative scope of CJ to formative assessment and professionalization.

Acknowledgement. This work was supported by Flanders Innovation & Entrepreneurship and the Research Foundation – Flanders (grant number 130043).

References

Andrich, D.: An index of person separation in latent trait theory, the traditional KR-20 index, and the Guttman scale response pattern. Educ. Res. Perspect. **9**(1), 95–104 (1982)

Anshel, M.H., Kang, M., Jubenville, C.: Sources of acute sport stress scale for sports officials: Rasch calibration. Psychol. Sport Exerc. **14**(3), 362–370 (2013)

Anseel, F., Lievens, F.: The mediating role of feedback acceptance in the relationship between feedback and attitudinal and performance outcomes. Int. J. Sel. Assess. **17**(4), 362–376 (2009). doi:10.1111/j.1468-2389.2009.00479.x

Bloxham, S.: Marking and moderation in the UK: false assumptions and wasted resources. Assess. Eval. High. Educ. **34**(2), 209–220 (2009)

Bradley, R.A., Terry, M.E.: Rank analysis of incomplete block designs, 1. The method of paired comparisons. Biometrika **39** (1952). doi:10.2307/2334029

Bramley, T.: Paired comparisons methods. In: Newton, P., Baird, J.-A., Goldstein, H., Patrick, H., Tymms, P. (eds.) Techniques for Monitoring the Comparability of Examination Standards, pp. 246–294. Qualification and Authority, London (2007)

Bramley, T. (ed.): Investigating the Reliability of Adaptive Comparative Judgment. Cambridge Assessment, Cambridge (2015)

Bouwer, R., Koster, M., Van den Bergh, H.: 'Well done, but add a title!' Feedback practices of elementary teachers and the relationship with text quality (2016) (Manuscript submitted for publication)

Collins, A., Joseph, D., Bielaczyc, K.: Design research: theoretical and methodological issues. J. Learn. Sci. **13**(1), 15–42 (2004)

Crisp, V.: Do assessors pay attention to appropriate features of student work when making assessment judgments? Paper presented at the International Association for Educational Assessment, Annual Conference, Baku, Azerbaijan (2007)

Hattie, J., Timperley, H.: The power of feedback. Rev. Educ. Res. **77**, 81–112 (2007)

Hevner, A.R.: A three cycle view of design science research. Scand. J. Inf. Syst. **19**(2), 87–92 (2007)

Kowalski, T.J., Lasley, T.J.: Part I: Theoretical and practical perspectives. Handbook of Data-Based Decision Making in Education, pp. 3–86. Routledge, New York (2009)

Laming, D.: Human Judgment: The Eye of the Beholder. Thomson Learning, London (2004)

Lesterhuis, M., Verhavert, S., Coertjens, L., Donche, V., De Maeyer, S.: Comparative judgement as a promising alternative. In: Cano, E., Ion, G. (eds.) Innovative Practices for Higher Education Assessment and Measurement. IGI Global, Hershey (2016, in Press)

Linacre, J., Wright, B.: Chi-square fit statistics. Rasch Measur. Trans. **8**(2), 350 (1994)

Luce, R.D.: Individual Choice Behaviors. A Theoretical Analysis. Wiley, New York (1959)

McMahon, S., Jones, I.: A comparative judgement approach to teacher assessment. Assessment in Education: Principles, Policy and Practice (ahead-of-print), pp. 1–22 (2014)

Mortier, A., Lesterhuis, M., Vlerick, P., De Maeyer, S.:Comparative Judgment Within Online Assessment: Exploring Students Feedback Reactions. In: Ras, E., Brinke, D.J.T. (ed.) Computer Assisted Assessment: Research into E-Assessment, CAA 2015. pp. 69–79. Springer-Verlag Berlin (2016)

Nicol, D.: Good design of written feedback for students. In: McKeachy Teaching Tips: Strategies, Research and Theory for College and University Teachers, Houghton Miffin, New York, 13th edn., pp. 108–124 (2009)

Nicol, D., MacFarlane-Dick, D.: Formative assessment and self-regulating learning: a model of seven principles of good feedback practice. Stud. High. Educ. **31**(2), 199–218 (2006)

Pierce, R., Chick, H.: Teachers' intentions to use national literacy and numeracy assessment data: a pilot study. Austr. Educ. Res. 38, 433–447 (2011)

Pollitt, A.: Let's stop marking exams. Paper presented at the annual conference of the International Association of Educational Assessment, Philadelphia, USA, 13–18 June (2004)

Pollitt, A.: The method of adaptive comparative judgment. Assess. Educ. Principles Policy Pract. **19**(3), 1–20 (2012). doi:10.1080/0969594X.2012.665354

Popham, J.: What's wrong - and what's right - with rubrics. Educ. Leadersh. **55**(2), 72–75 (1997)

Rasch, G.: Probabilistic Models for Some Intelligence and Attainment Tests (expanded edition). University of Chicago Press, Chicago (1960/1980). (Original work published in 1960)

Rossi, P.H., Freeman, H.E.: Evaluation: A Systematic Approach, 7th edn. Sage, London (2004)

Sadler, D.R.: Transforming holistic assessment and grading into a vehicle for complex learning. In: Joughin, G. (ed.) Assessment, Learning, and Judgment in Higher Education, pp. 1–19. Springer, Dordrecht (2009)

Schildkamp, K., Teddlie, C.: School performance feedback systems in the USA and in the netherlands: A comparison. Educ. Res. Eval. **14**(3), 255–282 (2008)

Shute, V.J.: Focus on formative feedback. Rev. Educ. Res. **78**(1), 153–189 (2008)

Thurstone, L.L.: The law of comparative judgment. Psychol. Rev. **34**(4), 273–286 (1927). doi: http://dx.doi.org/10.1037/h0070288

Vanhoof, J., Verhaeghe, G., Van Petegem, P.: Verschillen in het gebruik van schoolfeedback: een verkenning van verklaringsgronden. Tijdschrift voor onderwijsrecht en onderwijsbeleid 2008–2009(4) (2009)

Verhaeghe, G., Vanhoof, J., Valcke, M., Van Petegem, P.: Effecten van ondersteuning bij schoolfeedbackgebruik. Pedagogische Studiën **88**, 90–106 (2010)

Visscher, A.: School performance feedback systems. In: Visscher, A.J., Coe, R. (eds.) School Improvement through Performance Feedback, pp. 41–71. Swets & Zeitlinger, Lisse (2002)

van den Berg, I., Mehra, S., van Boxel, P., van der Hulst, J., Beijer, J., Riteco, A., van Andel, S.G.: Onderzoeksrapportage SURF-project: SCALA- Scaffolding Assessment for Learning (2014)

van der Hulst, J., van Boxel, P., Meeder, S.: Digitalizing feedback: reducing teachers' time investment while maintaining feedback quality. In: Ørngreen, R., Levinsen, K.T. (eds.) Proceedings of the 13th European Conference on e-Learning, ECEL-2014, Copenhagen, Denmark, pp. 243–250 (2014)

Xu, Y.: Examining the effects of digital feedback on student engagement and achievement. J. Educ. Comput. Res. **43**(3), 275–292 (2010)

Young, V.M.: Teachers' use of data: loose coupling, agenda setting, and team norms. Am. J. Edu. **112**, 521–548 (2006)

Student Teachers' Perceptions About an E-portfolio Enriched with Learning Analytics

Pihel Hunt[1(✉)], Äli Leijen[1], Gerli Silm[1], Liina Malva[1], and Marieke Van der Schaaf[2]

[1] University of Tartu, Tartu, Estonia
{pihel.hunt,ali.leijen,gerli.silm,liina.malva}@ut.ee
[2] Utrecht University, Utrecht, The Netherlands
M.F.vanderSchaaf@uu.nl

Abstract. In recent years the use of e-portfolios has increased in teacher education. Moreover, another rapidly evolving area in teacher education is learning analytics (LA). This paper reports the experiences of 13 student teachers in the implementation of an e-portfolio that is enriched with LA in a teacher education programme. Thirteen student teachers of primary school teacher curriculum received feedback and were assessed by their supervisors in their internship via an e-portfolio with LA. Questionnaire and a focus group interview were administered among the participants to identify how the student teachers perceive the job-fit and the effort expectancy of the e-portfolio with the LA application and to indicate the challenges encountered. The study revealed several positive evidences: the student teachers were on agreement that e-portfolio with LA is time-economising and easy to use. The student teachers appreciated that they received a good overview about their professional development. As a challenge, many student teachers questioned whether the use of e-portfolio can increase the quality of their professional activities (e.g. planning lessons, carrying out teaching activities and evaluating pupils' learning). Future research should focus on how to support the student teachers so that they would comprehend the benefits of the e-portfolio with LA on their professional activities and how to integrate human expertise in a more dynamic way.

Keywords: E-portfolio · Learning analytics · Teacher education

1 Introduction

1.1 Background

Feedback on and assessment of professional activities (e.g. planning and carrying out the learning activities, choosing and designing the appropriate learning materials, evaluating the pupils, etc.) in the workplace are critical in becoming a teacher [1, 2]. In teacher education often electronic portfolios (e-portfolios) are used to support workplace-based learning and assessment [3, 4]. Existing research shows several potential benefits of the use of e-portfolios in teacher education. Aligning the e-portfolio with teaching education programmes and the professional activities to be assessed gives the student teachers a possibility to reflect on their development and work and to improve their effectiveness as a teacher [5]. However, research also shows that the implementation of e-portfolios in teacher education may be a complex process [6]. Firstly, the use of e-portfolios requires a wide range of digital technology skills and proficiencies among the student teachers and their supervisors [7, 8]. In order to overcome this challenge, students need to be instructed and supported in the development of the necessary technology skills and understandings that guidance in terms of clearer expectations and requirements, more directions and modelling through examples, would make the e-portfolio more meaningful for them [9]. Secondly, one major challenge is the time and attention needed for working with the e-portfolio [10].

Furthermore, potential data in an e-portfolio assessment are underused. Therefore, in this article we will study the usefulness of a teacher e-portfolio that is enriched with Learning Analytics (LA). Learning analytics applications can be defined as the measurement, collection, analyses and reporting of data about learners and their contexts, for the purpose of understanding and optimising learning and the utilising of environments in which it occurs [11]. E-portfolios combined with LA can deal with the complexities of the workplace-based environment and can provide just-in-time high quality feedback to and assessment of student teachers.

In the implementation of new technology, it is important to study and understand the perceptions of the people using the technology. We use the Unified Theory of Acceptance and Use of Technology (UTAUT) which describes the significant determinants of user acceptance and usage behaviour [12]. More specifically we focus on two constructs: performance expectancy and effort expectancy. Performance expectancy is defined as the degree to which an individual believes that using the system will help him or her to attain gains in job performance [12]. Only the job-fit root construct about the extent to which an individual believes that using a technology can enhance the performance of his or her job [13] from this construct will be used. Effort expectancy is defined as the degree of ease associated with the use of the system. Two scales from this construct, complexity [13] and ease of use [14], will be used. Complexity describes the degree to which an innovation is perceived as relatively difficult to understand and use [13] and ease of use describes the degree to which an innovation is perceived as being difficult to use [14].

The aim of this study is to explore the perceptions of student teachers who used an e-portfolio that is enriched with LA during their internship. The specific research questions were the following:

1. How did the student teachers perceive the job-fit of the e-portfolio with the LA application?
2. How did the student teachers perceive the effort expectancy of the e-portfolio with the LA application?

1.2 Context of the Study: The E-portfolio and the LA Application

At first the most crucial activities that student teachers need to develop during their teacher education studies were developed [15]. The assessment rubric used in the current study targeted five professional roles and 12 professional activities teachers need to carry

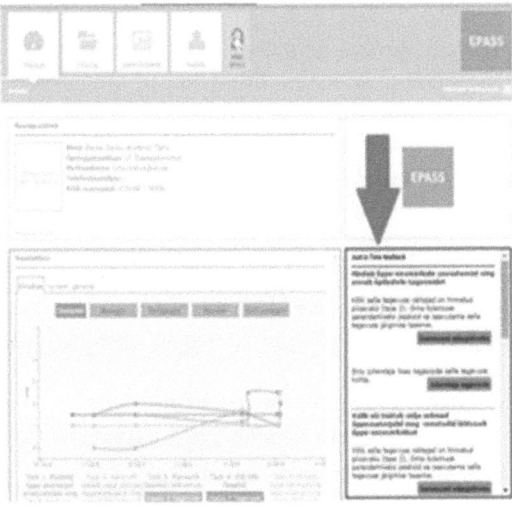

Fig. 1. The just-in-time feedback box on the dashboard of the e-portfolio.

out in their everyday work. Each activity was further specified in five performance levels (for further information about the assessment rubric see [16]).

While there was not a very clear system or protocol for providing feedback in the internship, there was a necessity for the implementation of a new system or protocol for formative assessment. Furthermore, there was a need for increasing student teachers' ownership for their learning and development as well. For these reasons, the above mentioned assessment rubric was implemented in an e-portfolio

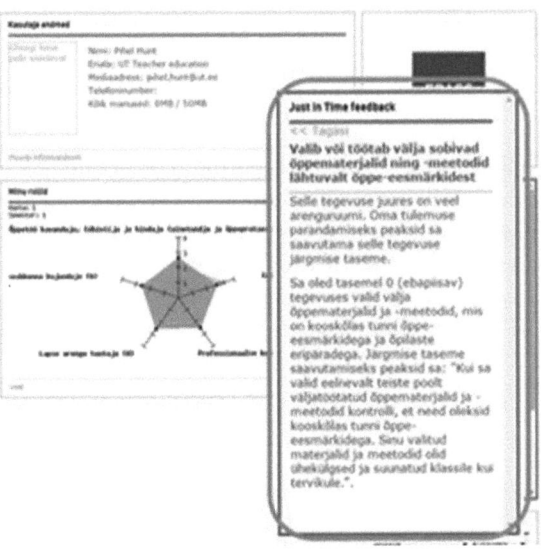

Fig. 2. Improvement feedback (automated feedback) in the just-in-time feedback box

environment. The rationale behind the e-portfolio was that student teachers should gradually submit their assessment data instead of submitting the data at the end of their internship. The student teachers filled in the electronic assessment forms (context info) in the e-portfolio and requested their supervisors to validate the forms. The supervisors marked the performance levels and provided further feedback. After the form was validated the student teachers received automated feedback in the automated feedback module. This appears on the e-portfolio dashboard in the Just-In-Time feedback box (see Fig. 1).

There are two kinds of feedback visible in the just-in-time feedback box - improvement feedback and supervisor feedback. This paper focuses on the improvement feedback, which is automated feedback based on the received rubric score given by the supervisor (Fig. 2).

In the following sections, we present our exploration of the perceptions of the student teachers in terms of the job-fit and effort expectancy of learning analytics enriched e-portfolio.

2 Method

2.1 Sample of the Participants

Data was collected from thirteen students of the final (fifth) year of the primary school teacher curriculum who used the assessment e-portfolio for six weeks during their internship. Nevertheless, using the e-portfolio was voluntary for them. This was agreed beforehand with the coordinator of the internship. All the student teachers were female and their age varied between 22 and 34 years. Mean age of the student teachers was 24.6 years ($SD = 3.3$). Most of the student teachers reported prior use of e-learning systems, mostly Moodle and eKool. No one had had any prior experiences with e-portfolios.

2.2 Procedure

Before student teachers started using the e-portfolio, the coordinator of the internship gave them an overview of the requirements of the internship. The student teachers were provided with manuals and videos on how to use the e-portfolio. Moreover, constant support was provided via Moodle environment (on the internship's page). Frequent reminders to use the e-portfolio were sent to the users via email and Moodle. Over the internship period of six weeks, the student teachers were asked to request assessment via the e-portfolio from their supervisors for eight lessons.

2.3 Instrumentation

The questionnaire was administered among the participants at the end of the internship. Based on the Unified Theory of Acceptance and Use of Technology [12] mentioned in the introduction section, the questions focused on the constructs of job-fit and performance expectancy. The specific questions that were used are presented in Table 1.

Table 1. Constructs, items and cronbach's alphas

Construct	Items[a]	Cronbach's alpha (e-portfolio; automated feedback module)
Job-fit [13]	Use of the system increases the quality of the professional activities I need to carry out during my internship Use of the system results in higher performance in the professional activities I have to carry out during my internship Use of the system supports the development of my professional activities during my internship In conclusion, the system helps to do my professional activities better	.94; .81
Effort expectancy ([12] adapted from [13, 14])	Use of the system means I have to do too much tedious work (e.g., data input, data retrieval) It takes a long time to learn how to use the system Using the system takes much of the time meant for carrying out my professional activities The system is easy to use	.70; .82

[a]The formulations of the items were adapted regarding the topic (e-portfolio, automated feedback module)

All questions were rated on a 5-point Likert scale (1-fully disagree; 5-fully agree). There was an option to answer "not applicable" (NA) if there was a possibility that the respondent had not used some parts of the system. Also, student teachers had the opportunity (open-end question) to add additional comments about the e-portfolio and the automated feedback module.

In addition to the questionnaires a focus group interview was conducted with three student teachers to get more information in order to explain more the results from the questionnaire and find the issues that may not have revealed from the questionnaire. The student teachers gave their permission to use their data from the e-portfolio, the questionnaire and the interview. They were granted confidentiality.

2.4 Data Analysis

The data was analysed in two steps. Firstly, descriptive statistics were used to describe the mean scores and the standard deviations (SD) of the constructs. The reliability analysis (Cronbach alphas and Spearman correlation coefficients) was carried out to analyse the questions that were answered on Likert scale. Also, Wilcoxon signed ranks test was used to evaluate significant differences between constructs. Secondly, a qualitative analysis of the open-ended responses and of the interviews was conducted. The qualitative analysis followed the thematic analysis procedure [17] in which the researcher looks for themes and sub-themes underlying the qualitative data.

3 Results

3.1 Student Teachers' Job-Fit of the E-portfolio and Automated Feedback Module

The mean score of the job-fit of the e-portfolio was 2.7 (SD = .83; N = 13). This result indicates that few student teachers agreed that the use of the e-portfolio resulted in increased quality of their professional activities, in higher performance in the professional activities or supported their development in their professional activities. Only two student teachers agreed that using the e-portfolio helped to do their professional activities better during the internship.

The mean score of the job-fit of the automated feedback module was 3.0 (SD = .92; N = 10). This score is a bit higher than the mean score for the e-portfolio although, there is no significant difference between these scores. There is a strong significant positive correlation (ρ = .81) between the job-fit of the e-portfolio and the automated feedback module. However, it must be also noted, that only 10 respondents out of 13 answered these questions. This may refer to the fact, that some of the respondents did not use the automated feedback module.

The student teachers were more positive in the interview. From the interview and open-ended responses it can be included that overall, student teachers were satisfied with their experiences in using the e-portfolio in their internship. They expressed that although they were a little sceptical before using it, they liked the idea that their professional development was easily observable.

3.2 Student Teachers' Effort Expectancy of the E-portfolio and the Automated Feedback Module

The mean score of the effort expectancy of the e-portfolio was 3.4 (SD = .74, N = 13). The opinions about whether using the e-portfolio means too much tedious work or not were divided into two: seven people agreed that it does and six disagreed. Most of the student teachers agreed that the e-portfolio was easy to use and did not take too much of their time meant for professional activities. A common view among the participants was that the positive side of the implementation of the e-portfolio would be the decrease in the use of paper. Some of the student teachers indicated that they experienced some technical problems with the system. These problems were mostly connected with logging in to the system or sending the feedback protocol to the supervisor.

The mean score of the effort expectancy of the automated feedback module was 3.4 (SD = .85; N = 10). The Spearman correlation analysis shows a significant positive correlation (ρ = .76) between the effort expectancy of the e-portfolio and effort expectancy of the automated feedback module and there is no significant difference between the scores. Similarly to the job-fit construct, three participants did not answer to the questions about the automated feedback module, which may suggest the fact that the student teachers did not use this module.

The most interesting topic that emerged from the open-ended questions and from the interview was the perception that the automated feedback was not personalised because it was created based on the rubric and on the supervisor's score, so every student teacher could receive the same feedback. The student teachers pointed out that they value more the oral feedback or the written comments they received from their supervisors rather than the automated feedback generated by the computer.

4 Main Conclusions and Implications

The aim of this study was to explore the job-fit and the effort expectancy of an e-portfolio with the LA application which was used by thirteen student teachers during their internship. This study provided useful insights into benefits and challenges of implementing e-portfolio and LA in educational contexts and therefore provides a valuable base for implementations and further developments.

The study revealed several positive evidences regarding the effort expectancy of the e-portfolio and automated feedback. Student teachers were on agreement that e-portfolio with LA is time-economising and easy to use. This finding is contrary to previous studies [7, 8, 10] which have suggested that working with e-portfolios requires too many technical skills, and too much time and attention. The student teachers expressed that although they were a little sceptical before using the e-portfolio, they liked the idea that their professional development was easily observable. A common view among the participants was that the positive side of the implementation of the e-portfolio would be the decrease in the use of paper.

However, the study also revealed that student teachers only modestly value the e-portfolio and automated feedback as a beneficial tool for their professional development. Many student teachers questioned whether the use of e-portfolio can increase the quality of their professional activities (e.g. planning lessons, carrying out teaching activities and evaluating pupils' learning). A possible explanation for this might be that for gaining more from the e-portfolio, student teachers should reflect on their development and work. However, the e-portfolio does not automatically lead to reflection; it only works when it is well embedded in the educational environment and the e-portfolio is aligned with learning goals of the teacher education programme, assignments, assessment of the internship and instruction. Moreover, in the era of complex ICT-enriched environments, participants seem to value expert insights and communication of humans over mere computer-based automatic output.

The major limitation of this study is the small sample size, making it difficult to generalize the results to a larger population. Future research should focus on how to support the student teachers so that they would comprehend the benefits of the e-portfolio with LA on their professional activities and how to integrate human expertise in a more dynamic way.

References

1. Ericsson, K.A., Charness, N., Feltovich, P.J., Hoffman, R.R. (eds.): The Cambridge Handbook of Expertise and Expert Performance. Cambridge University Press, New York (2006)
2. Hattie, J.: Visible Learning. A Synthesis of Over 800 Meta-analysis Relating to Achievement. Routledge, Oxon (2009)
3. Van der Schaaf, M., Stokking, K., Verloop, N.: Teacher beliefs and teacher behaviour in portfolio assessment. Teach. Teach. Educ. **24**(7), 1691–1704 (2008)
4. Bok, G.J., Teunissen, P.W., Spruijt, A., Fokkema, J.P.I., van Beukelen, P., Jaarsma, A.D.C., van der Vleuten, C.P.M.: Clarifying students' feedback seeking behaviour in clinical clerkships. Med. Educ. **47**(3), 282–291 (2013)
5. Parker, M., Ndoye, A., Ritzhaupt, A.D.: Qualitative analysis of student perceptions of E-portfolios in a teacher education program. J. Digit. Learn. Teach. Educ. **28**(3), 99–107 (2012)
6. Granberg, C.: E-portfolios in teacher education 2002–2009: the social construction of discourse, design and dissemination. Eur. J. Teach. Educ. **33**(3), 309–322 (2010)
7. Oakley, G., Pegrum, M., Johnston, S.: Introducing E-portfolios to pre-service teachers as tools for reflection and growth: lessons learnt. Asia Pac. J. Teach. Educ. **42**(1), 36–50 (2014)
8. Wray, S.: E-portfolios in a teacher education program. E-learning **4**(1), 40–51 (2007)
9. Woodward, H., Nanlohy, P.: Digital portfolios in pre-service teacher education. Assess. Educ. Princ. Policy Pract. **11**(2), 167–178 (2004)
10. Evans, M.A., Powell, A.: Conceptual and practical issues related to the design for and sustainability of communities of practice: the case of E-portfolio use in preservice teacher training. Technol. Pedagogy Educ. **16**(2), 199–214 (2007)
11. Solar: Open Learning Analytics: an integrated & modularized platform (2011). http://www.solaresearch.org/wp-content/uploads/2011/12/OpenLearningAnalytics.pdf
12. Venkatesh, V., Morris, M.G., Davis, G.B., Davis, F.D.: User acceptance of information technology: toward a unified view. MIS Q. **27**(3), 425–478 (2003)
13. Thompson, R.L., Higgins, C.A., Howell, J.M.: Personal computing: toward a conceptual model of utilization. MIS Q. **15**(1), 124–143 (1991)
14. Moore, G.C., Benbasat, I.: Development of an instrument to measure the perceptions of adopting an information technology innovation. Inf. Syst. Res. **2**(3), 192–222 (1991)
15. Krull, E., Leijen, Ä.: Perspectives for defining student teacher performance-based teaching skills indicators to provide formative feedback through learning analytics. Creat. Educ. **6**(10), 914–926 (2015)
16. Leijen, Ä., Slof, B., Malva, L., Hunt, P., van Tartwijk, J., van der Schaaf, M.: Performance-based competency requirements for student teachers and how to assess them. Int. J. Inf. Educ. Technol. **7**(3), 190–194 (2017)
17. Braun, V., Clarke, V.: Using thematic analysis in psychology. Qual. Res. Psychol. **3**(2), 77–101 (2006)

Support in Assessment of Prior Learning: Personal or Online?

Desirée Joosten-ten Brinke[1,2(✉)], Dominique Sluijsmans[3,4], and Wim Jochems[1,5]

[1] Welten Institute, Open Universiteit of the Netherlands,
Postbus 2960, 6401 DL Heerlen, The Netherlands
Desiree.joosten-tenbrinke@ou.nl
[2] Fontys University of Applied Sciences,
P.O. Box 90900, 5000 GA Tilburg, The Netherlands
[3] Zuyd University of Applied Sciences,
P.O. Box 550, 6400 AN Heerlen, The Netherlands
[4] Maastricht University, P.O. Box 616,
6200 MD Maastricht, The Netherlands
[5] Jochems Advies and Review, P.O. Box 90900,
6371 CC Landgraaf, The Netherlands

Abstract. Assessment of Prior Learning (APL) offers significant benefits to adult learners. It reduces the gap between educational programmes and the labour market and provides learners the possibility to shorten their prospective study programmes. APL however requires adequate support of the learners that engage in APL. This study investigates which support possibilities could be useful and efficient for APL candidates in a distance education context. Staff members of an APL provider (APL tutors and educational scientists), APL candidates and a group of experts on online support evaluated and discussed the possibilities of both embedded and personal support in APL.

With regard to the different phases that can be distinguished in APL, the results show that all participants expect support particularly in the phase of gathering appropriate evidence. From the staff members' perspective, embedded support is most appropriate and many types of support provisions are recommended. APL candidates prefer a combination of embedded and personal support, whereby the type of personal support can be limited to telephone and email support. The latter is preferred because of its personal, to-the-point and time independent character. An overview of the highest added value of support as well as support efficiency is provided. Unfortunately, the highest added value is not always the most efficient. Guidelines for the elaboration of efficient support with high added value in APL in the context of distance education are presented.

Keywords: Assessment of prior learning · Embedded support · Personal support

1 Introduction

Lifelong learning is an essential concept in higher education. It is seen as an opportunity to meet the needs of individuals, educational institutes and employers (Corradi et al. 2006; Miguel et al. 2015; Sanseau and Ansart 2014) and it promotes forms of education that are not primarily aimed at acquiring diplomas, but also to education that recognizes competences that have been gained elsewhere and in alternative ways. The life experiences of adult, often working learners, who are experienced, motivated persons who take responsibility for their learning (Knowles et al. 2014), are worthy of recognition when they enrol for a study programme. Education for working adults requires other support mechanisms and other curricula than full time education (Committee flexible higher education for labour 2014; SER 2012). Next to formal learning, they acquire competences through non-formal learning (i.e., intentional learning within a structured context, but without legally or socially recorded certification) and informal learning (i.e., not intentional, not structured and does not lead to certification). Because of these prior learning experiences, specific entry routes are necessary (Schuetze and Slowey 2002).

Many terms exist to define procedures in which prior learning is recognised (Joosten-ten Brinke et al. 2008). In this study, the term 'Assessment of Prior Learning' (henceforth abbreviated as APL) is used, whereby the prior learning contains informal and non-formal learning. APL is seen as a procedure that acknowledges that the individual's self-concept and life experiences constitute an avenue for further learning (Cretchley and Castle 2001). In general, APL consists of four phases (see e.g., The Calibre group of Companies 2003; Thomas et al. 2000; Vanhoren 2002). In the first, learner profiling phase, the institute provides information about APL possibilities and its procedure. In the second phase of evidence-gathering, learners collect evidence about previous experience to support a claim for credit with respect to the qualification they want to achieve. In the third, assessment phase, assessors review the quality of the learner's evidence using set assessment standards. The final phase, recognition, involves verification of the assessment outcome through, for example, the issuing of credits. Learners intending to start an educational programme then receive a study advice.

APL offers significant benefits to adult learners: it provides a better connection between educational programmes and the labour market, emphasises lifelong and flexible learning, and increases efficiency for part-time adult learners by shortening their programmes and reducing course loads and costs (Wihak 2007). Despite the many benefits and the extensive use of a credit framework, a part of the universities is not always motivated to implement APL (Kawalilak and Wihak 2013). A reason for this attitude can be found in the universities' concern that the intellectual foundation of university learning is threatened by APL (Thomas et al. 2000; Wihak 2007). The role of learning in informal and non-formal learning environments was more or less denied, the introduction of APL is expensive because of the development of policy, instruments and training and is seen as time-consuming (Thomas et al. 2000; Wheelahan et al. 2002). Finally, there is a concern that learners need too much support in gathering the appropriate evidence (Scholten et al. 2003; Thomas et al. 2000; Wheelahan et al. 2002). With regard to this concern, candidates indeed find it difficult to give adequate descriptions of former learning experiences (Shapiro 2003). Three reasons are given. Firstly, learners do not always realise the extent of their knowledge and competences, and might lack the appropriate language to articulate them, that is, they need support translating their knowledge and skills into educational discourse. An additional problem here is that the educational discourse not always is described as clear as it should be. In some cases, candidates are asked to define their mastery level on competences, while the educational programme is not competence-based at all. Secondly, if learners realise that they have learned a lot, there are circumstances where the choice of APL is not at all advisable. For example, if exemptions, as an outcome of APL, are not accepted in another context. Thirdly, the perception of informal learning in particular is subjective, which makes it difficult for APL candidates to assess whether prior experiences have actually contributed to their learning (Colley et al. 2002; Wheelahan et al. 2002). Scheltema (2002) also showed that candidates know what is expected of them, but need more support in translating the collected material into appropriate evidence. They need help reflecting on their own competences and preparing a competence profile, gathering the appropriate evidence and composing the portfolio (Scholten et al. 2003; Dutch Ministry of Economic Affairs 2000). One-step in improving the adoption of APL in universities, is making the support as efficient as possible for the institutes and as valuable as possible for the candidates.

Although more insight is needed in ways to support learners in APL, we found that little research is available as to how learners could best be supported in APL, in terms of appreciation (high valued support) and efficient support. In the context of open and distance learning, Tait (2000) defines support as '… the range of services […] which complement the course materials or learning resources' (p. 289). Functions of support are administration, guidance, direction, advice and encouragement administration, encouragement, and feedback (Jacklin and Robinson 2007; Macdonald and McAteer 2003). Day (2001) refers to specific functions of APL support: it should help candidates identify relevant learning, make action plans to demonstrate this learning, and prepare and present evidence for assessment.

Support can be provided in different ways. A distinction can be made between embedded and personal support. Embedded support consists of techniques incorporated in printed or technology enhanced materials (Martens and Valcke 1995). Examples are worked-out examples, online fora, FAQs or elaborated criteria. Macdonald and McAteer (2003) describe personal support as that given by a person (e.g., tutor or study adviser) in real time or asynchronously either in a group (e.g., face-to-face tutorials) or individual context (telephone, email, etc.). Donoghue et al. (2002) describe assistance strategies including writing skills workshops, library orientations, acknowledgment of learners' classroom needs, discussions with staff, critical thinking and analysis, literature searches, application of literature findings, development of a position, and use of argument and referencing procedures. Macdonald and McAteer (2003) have evaluated the potential value of combining the available provisions to enhance learner support for distance and campus-based universities, and stress the importance of creating a balance between embedded support, such as online conferences, and personal tutor support. This combination of support types has proven to be helpful to learners (Mason 2003).

In this study, we are interested in the perception of both staff and candidates in APL procedures about valuable and efficient support. Therefore, the presented study focuses on an inventory of support possibilities in each APL phase in the context of higher distance education. Three questions are central:

(1) In which APL phase is embedded and/or personal support desired?
(2) Which specific types of personal and embedded support and support provisions are most desired in APL?
(3) Which type of support has the highest added value and is the most efficient?

2 Method

Context

At a distance university for lifelong learning in the Netherlands, learners can receive exemptions for parts of the curriculum if they already have completed formal higher education. In addition, an APL procedure was available in which learners receive credit points if their informally or non-formally acquired competences match the learning outcomes of an accredited educational programme. Based on a study on the candidates' expectations of needed support provisions (Joosten-ten Brinke 2007), the support for each phase was organised as follows:

(1) In the candidate profiling phase, embedded support consisted of a standardised email providing basic information including the web and email address of a general tutor, a website with general information about the procedure, standard (work experience) requirements for the different phases, a manual and a sample portfolio format. Personal support consisted of individual email, telephone, and face-to-face sessions on request.
(2) In the evidence-gathering phase, embedded support consisted of a portfolio format, a standardised example of argumentation and the educational programme's learning objectives. Personal support comprised email, telephone, face-to-face sessions on request and portfolio feedback.
(3) In the assessment phase, embedded support consisted of standardised invitations for (parts of) the assessment. No personal support was provided.
(4) In the recognition phase, embedded support consisted of a standardised letter of recognition. Personal support was supplied by way of personal study advice on request.

Participants

Three groups of participants were involved: staff members, an expert group and an APL candidates' group. The first group consisted of seven tutors and three educational scientists. These staff members had between five and 20 years of experience in supporting learners and the credit exchange for formal learning in open and distance learning, knowledge of APL and some experience (one year) in APL support. The second group comprised three experts in online support in higher education domains. This expert group reviewed the first results of the staff members and provided an overview of different support provisions. The third group consisted of eight APL candidates (5 men, 3 women) distributed over five domains: Educational Science

Table 1. Focus group session interview scheme

Personal support
I. In which APL phases should personal support be available?
i. What personal support do you see for each phase?
1. Which personal support has the highest added value for the candidate?
2. Which personal support is the most efficient?
Embedded support
II. In which APL phases should embedded support be available?
i. What embedded support possibilities do you see for each phase?
1. Which embedded support has the highest added value for the candidate?
2. Which embedded support is the most efficient?

($n = 2$), Law ($n = 1$), Management Science ($n = 2$), Psychology ($n = 1$) and Computer Science ($n = 2$). Five candidates completed the APL procedure successfully, one returned a negative result and two did not complete the procedure.

Materials

A focus group session with the staff members aimed to identify the types of APL support desired by the institute. A question scheme for this session is presented in Table 1. An electronic Group Support System (eGSS), or computer-based information processing system designed to facilitate group processes, was used to support the session. This allows collaborative and individual activities such as the brainstorming, sorting, rating and clustering of ideas via computer communication. All staff members (n = 10) are seated in front of a laptop connected to a network and facilitator computer. Input generated from the expert session was collected and saved in the eGSS, which delivers anonymous results.

The online support experts received the results of the focus group session and were asked to analyse and evaluate the results of the staff members based on their expertise on online support tools. An open interview was used for a group evaluation. Finally, a structured interview scheme was used for the individual candidate interviews. This scheme was similar to that used in the focus group session, with the exception that the results of the focus group session were included. Candidates were sent the interview questions beforehand for preparation. The interviews were audiotaped.

Procedure

At the start of the focus group session, the researcher gave broad definitions of both APL and support. In accordance with the interview scheme (see Table 1), the staff members of the university who participated as a focus group were then asked to vote within five minutes for the phase (candidate profiling, evidence gathering, assessment and recognition) in which support was most desired. Next, they had seven minutes to think individually about different types of support, which were gathered by the eGSS and listed (anonymously) on the screen. They were then asked to review this list for each APL phase and to combine or erase duplicates. On the remaining list, they had to indicate the two types of support with the highest added value, and the two with the greatest efficiency. A moderator leaded this focus group session and discussed the given answers after each question. Answers with the same meaning were than combined after agreement of the individual members.

One week after the focus group session, the results of this session were discussed in person by a review group of three online support experts, who were instructed to make relevant additions to the list of support options.

Two weeks after the focus group session, APL candidates were interviewed using a structured interview scheme to check whether the support provisions suggested by the institute matched those desired by APL candidates.

Data Analysis

First, the focus group session is analysed. Because the contribution of each staff member in the focus group was anonymous, it is not possible to trace the results to individuals. Therefore, these analyses are descriptive in terms of percentages and qualitative overviews. The first research question ('In which APL phase is embedded and/or personal support desired?') was analysed by way of the percentage of desired support in each phase. To answer the second question ('Which specific types of personal and embedded support and support provisions are most desired in APL?'), the mentioned embedded and personal support provisions were listed. Subsequently, a list of support topics for APL mentioned by the staff members was generated. To answer the third research question ('Which type of support has the highest added value and is the most efficient?'), the staff members' votes were listed. To come to a high valued and efficient support framework, a selection is made of the support possibilities with a minimum score of 2 on both highest added value and efficiency. After the analyses of the staff members' data, the results on the second and third question were supplemented with the results of the discussion with the online support experts. The interviews with the candidates were first analysed separately, and then added to the results of the prior two groups.

3 Findings

This study focuses on the support possibilities for APL candidates in distance higher education from the perspective of an APL provider and APL candidates. The results are presented on the basis of the interview questions.

In which APL phase is embedded and/or personal support desired?

Table 2 gives an overview of the percentage of institute participants and candidates who desire support in the different APL phases. In the evidence-gathering phase, all candidates desire both personal and embedded support. In contrast, only 70% of the staff members desire personal support in this phase, and 80% were in favour of

Table 2. Phases with support desired by institute and candidates indicated as percentages.

APL phase	Personal support		Embedded support	
	Institute	Candidates	Institute	Candidates
1. Candidate profiling	40%	66%	100%	66%
2. Evidence gathering	70%	100%	80%	100%
3. Assessment	40%	33%	50%	0%
4. Recognition	10%	16%	50%	0%

embedded support. In the assessment phase, no candidate desires embedded support; in the recognition phase, staff members and candidates have low demands for support.

Table 3 classifies the appropriate support provisions, which can be divided into (1) provisions for embedded (electronic or written) and personal support; and (2) those that support the individual versus the group. Each group named the website, FAQs, manuals and printed materials as appropriate types of embedded support. They all mentioned emails to individuals, face-to-face contact, telephone conversations and a general information session as relevant personal support. The expert group suggested more topical provisions such as virtual classrooms, telephone conferences, mailing lists and video conferencing.

Table 3. Appropriate support provisions according to institute, expert group and APL candidates

		Support provision	Institute	Expert group	Candidates
Embedded support	Technology enhanced	Computer system	•	•	
		Online self-assessment to test whether APL procedure could be worthwhile	•		
		Search engine	•	•	
		Automatic alerts	•	•	
		Automatic email	•	•	
		Video (interviews with former APL candidates)	•	•	
		Website	•	•	•
		FAQs	•	•	•
	Written	Learner newspaper	•	•	
		Leaflet (APL manual; study guide)	•	•	•
		Portfolio format	•		•
		Printed examples of good and bad practice	•	•	•
		Feedback		•	
		Jurisprudence for similar cases	•		
		Letter to individuals		•	
Personal support	Individual	Email	•	•	•
		Face-to-face	•	•	•
		Telephone	•	•	•
	Group	Workshop	•	•	
		General information session	•	•	•
		In-company training	•	•	
		Virtual classroom/computer- and video conference		•	
		Telephone conference		•	
		Email lists		•	
		Letters		•	

Table 4 shows the support topics which can be given in each phase, including both embedded and personal support.

Table 4. Overview of topics in each phase for embedded or personal support

Topics for support	Embedded support	Personal support
1. Candidate profiling		
Whole procedure	•	
All information sources		•
Educational programme standards/competences	•	
Portfolio structure		•
Lack of clarity in embedded support		•
2. Evidence gathering		
Possible standard outcome		•
Examples of evidence	•	•
Composition of portfolio (in view of assessment criteria)	•	•
Overview of competences	•	•
APL procedure		•
CV suggestions		•
Employer's certificate	•	•
Analogous cases	•	
Standardised CV	•	
Types of evidence	•	•
3. Assessment		
Procedure	•	
Criteria and standards	•	
Information about assessment phase	•	
Strategies in assessing competencies	•	•
Assessment possibilities		•
Protocol	•	
Good and bad portfolios for competence assessment	•	
Former assessment results (jurisprudence)	•	•
Assessment phase in view of criteria		•
4. Recognition		
Standard recognitions		•
Phase procedure	•	
Explanation of (possible) recognition result	•	•
Competences lacking		•
Civil effect		•
Comparable results (jurisprudence)	•	•
Alternative studies available for recognition		•
Recognition opportunities		•
Alternatives to continuation/study advice	•	•
Relevant studies		•
Complementary activities		•
Upon negative result		•

Table 5. Overview of highest added-value support and scores for efficiency

Support possibilities	Embedded support		Personal support	
	Highest value[a]	Efficiency[a]	Highest value[a]	Efficiency[a]
1. Candidate profiling				
General information session			4	2
Initial individual face-to-face conversation			3	0
Standard individual face-to-face conversation			3	2
Information by telephone			2	0
Meeting with former APL candidates			2	2
List of information sources (links, websites) by email			0	5
APL manual	5	4		
Self-assessment instrument to test if APL procedure could be worthwhile	5	3		
FAQs	4	6		
Website	4	4		
Good and bad examples with clarification	3	2		
Interviews with former APL candidates	1	2		
Portfolio format	1	1		
Jurisprudence for similar cases	1	1		
Overview of competences per educational programme	1	0		
Study guide/flyer	0	4		
Standards	0	1		
2. Evidence gathering				
Individual support for composing portfolio			4	1
Written comments on portfolio			4	1
Discussion of portfolio in view of assessment criteria			4	0
Workshop on how to compose portfolio			2	2
CV suggestions			2	1
Discussion of evidence examples			2	4
Answering of questions by phone or email			1	4
Mind-manager system with portfolio format	9	7		
Good and bad examples with clarification	8	5		

(*continued*)

Table 5. (*continued*)

Support possibilities	Embedded support		Personal support	
	Highest value[a]	Efficiency[a]	Highest value[a]	Efficiency[a]
Manual on how to compose portfolio	4	2		
Electronic seeking and presenting of analogous cases	3	2		
Standardised CV	2	2		
FAQs	1	4		
E-portfolio format	1	3		
List with evidence examples	1	1		
Overview of assessment criteria	0	3		
Email alerts	0	1		
3. Assessment				
Individual face-to-face conversation re. assessment criteria			6	3
Discussion of former assessment results			3	3
Assessment criteria	9	6		
Elaboration of assessment protocol	7	2		
Good and bad examples with clarification of portfolios	5	4		
Assessment results (jurisprudence)	2	5		
Contact with other candidates with same question	1	0		
4. Recognition				
Availability for individual explanation (after recognition)			6	1
Answering of questions by email			4	4
Group discussion to compare results			2	1
Referral to others			1	4
Examples of cases in which recognition was and was not given	6	3		
Standard recognitions	4	6		
Phase procedure	4	4		
Graphic overviews of recognisable programme elements	4	2		

[a]Maximum score is 10 (=number of staff members in the focus group).

Which type of support has the highest added value and is the most efficient?

The first column of Table 5 presents an overview of possible support for each APL phase, comprising combinations of support types and provisions per phase as given by the staff members. In the second and third columns, the scores are given for the highest added value and efficiency of personal support, while those for embedded support are shown in the fourth and fifth columns. The maximum score is 10 (=number of staff members in the focus group session).

Seven out of the eight APL candidates described the combination of personal and embedded support as having the highest added value. The most efficient method, it appears, is to first read all the information supplied on the website and in manuals (embedded support) and then, if necessary, ask any remaining questions by email. The preference for email is explained by the opportunity to formulate one's question/s adequately, receive a written answer and do so time-independently.

The expert group believes a virtual classroom to be a useful instrument. This is a private, online space offering all the activities that tutors might use to support APL candidates in a real-world classroom. The context of distance education is ideal for this provision. It offers a solution to efficiency problems, such as travel time, and gives the possibility to bring together candidates from all over the country.

4 Conclusion and Discussion

I Little research evidence is at hand how learners who participate in an APL procedure can be best supported in gathering relevant evidence of prior learning. Due to this gap in APL research, we made a first attempt in understanding which support is desired by the relevant stakeholders in APL and how efficient and valuable this support is perceived. For this, we investigated the desired support provisions in each phase of the procedure, and the provisions with the highest added value and efficiency. According to the staff members it is possible to provide support in all phases of APL, though embedded support is seen as more appropriate than personal support. Especially, face-to-face personal support causes problems in distance education context. Still, all candidates showed interest in personal as well as embedded support in the evidence-gathering phase. The difference between the institute's desired embedded support and that of the candidates in the assessment and recognition phases is remarkable: candidates did not expect embedded support in these two phases. Candidate interviews showed, however, that they appreciate personal support after the recognition phase, for example in the form of study advice. The second, evidence-gathering phase is mentioned most; this is in line with earlier research (Colley et al. 2002; Cretchley and Castle 2001; Shapiro 2003; Spencer et al. 2000; Wheelahan et al. 2002). The use of technology seems to be more important for the institute than for the candidate. This might be a consequence of the objective of the APL procedure. The main goal for the candidate of APL is more content related in receiving some exemptions for his future learning path; for the institute it is to support this trajectory as efficient as possible.

The results to the second question revealed that staff members see more possibilities for support than the APL candidates desire or are familiar with. This result joins the finding of Martens et al. (2007) that positive expectations of developers, in their research in the context of constructivist e-learning environments, do not always match the actual perceptions of learners. Candidates prefer email support after exhausting avenues for embedded support on the website and in manuals, and gave the following reasons for this preference: it can be given personally, and is to-the-point and time independent. The expert group identified a virtual classroom as a useful instrument, especially in the context of distance education. However, Gillies (2008) mentions logistic problems with this kind of support provisions, because it requires that distant learners have to be present at a set time in a place where the support provision is available. The combination of embedded support and the availability of personal support has, according to the candidates, the highest added value.

Regarding the third research question, Table 5 provided an overview of the highest added-value and efficient support possibilities. Unfortunately, but not surprisingly, the highest added-value support is not always seen as the most efficient. For example, discussing the candidates' portfolios according to the assessment criteria was seen as having high added value by four staff members, yet none of them found it efficient. The question thus arises as to how to make this part more efficient with the use of technology. Portfolio discussions, for example, can be held in a virtual classroom or through group sessions, both of which are more efficient than individual, face-to-face sessions. Moreover, personal support is less efficient for the institute than embedded support; it is therefore important to embed support as far as possible. In some cases, however, the development of embedded support will mean high investment costs. One must then weigh the extent to which the result would be worthy – in other words, the added value. For example, the mind-manager system within the portfolio format scores highly on both added value and efficiency. The value of implementing it, then, is obvious. However, if such a system is not available, its purchase or development must be taken into account. To decrease the time required of the institute, it is also necessary to identify the functions which can be conducted by computer, such as online self-assessment of prior learning. A self-assessment instrument for APL in distance education is useful in deciding whether APL will be meaningful for candidates or not (Joosten-ten Brinke et al. 2009).

Based on the results presented in Table 6 and the candidates' perceptions, we propose a new support framework. We have selected support possibilities with a minimum score of 2 on both highest added value and efficiency. This would imply that in the candidate-profiling phase, general information sessions could be held once or twice a year with the possibility for face-to-face conversation. Embedded support in this phase should consist of at least an APL manual, an online self-assessment instrument to test whether the procedure is likely to be meaningful, a website with FAQs and information about APL, and finally good and bad examples of portfolios with clarification.

In the second APL phase, evidence-gathering, personal support should preferably consist of discussions about examples of evidence. Embedded support can include a mind-manager system with a portfolio format and strong versus poor examples; a 'how to compose a portfolio' manual; the opportunity to electronically seek and present analogous cases, and a FAQs list. In addition, instead of a face-to-face workshop on portfolio composition, we propose using a more efficient virtual classroom workshop. This could be offered a few times a year, with candidates from all over the country applying by email. Further research is needed on the possibilities for automated feedback on portfolios (Denton et al. 2008). The advantages is that automated feedback can be provided quickly and directs the candidate for continuing or change further steps.

Table 6. Comparison of existing and proposed frameworks for APL support

Existing framework		Proposed framework	
Embedded support	Personal support	Embedded support	Personal support
1. Candidate profiling			
Standardised email with basic information and reference	Email	Self-assessment instrument	General information session once or twice a year
Website with general information about procedure and general requirements	Telephone	Website with all APL information	Voluntary face-to-face standard conversations
APL manual	Face-to-face sessions on request	APL manual	Personal email for interim questions
Portfolio format		Good and bad examples with clarification	
		FAQs	
2. Evidence-gathering			
Portfolio format	Email	Mind manager system with portfolio format	Discussions about evidence examples
Standardised, elaborated example of argumentation	Telephone	Good and bad examples with clarification	Workshop by virtual classroom
Learning objectives of educational programme	Face-to-face session on request	Manual: How to compose a portfolio	Personal email for interim questions
	Portfolio feedback	Electronic seeking and presenting of analogous cases	
		FAQs	
		Standardized feedback	
3. Assessment			
Standardised invitations for (parts of) assessment	No personal support	List of criteria	Individual face-to-face conversation based on assessment criteria
		Elaboration of protocol	Discussion about former assessment results
		Good and bad portfolio examples	Personal email for interim questions

(*continued*)

Table 6. (*continued*)

Existing framework		Proposed framework	
Embedded support	Personal support	Embedded support	Personal support
		for competence assessment	
		Automated scoring (LSA)	
		Overview of jurisprudence on assessment results	
4. Recognition			
Standardised letter for recognition	Study advice on request	Examples of cases in which recognition was and was not given	Personal email for interim questions
		Description of standard recognitions and recognition phase	
		Graphic overviews of the recognisable programme elements	

In the assessment phase, personal support should comprise an individual, face-to-face conversation based on the assessment criteria and former assessment results. Embedded support should consist of a list of assessment criteria, an elaboration of assessment protocol, examples of good and bad portfolios for competence assessment, and an overview of assessment results jurisprudence. For the support of the staff, the help of automated scoring techniques (for example Latent semantic analysis (LSA) for positioning candidates should be helpful (see Van Bruggen et al. 2009). Finally, in the recognition phase, questions can be answered by way of personal emails; embedded support should include examples in which recognition was and was not given, descriptions of standard recognitions and the phase itself, and graphic overviews of the educational programme.

Based on the comments of APL candidates, we suggest providing them with the email address of a tutor available to deal with interim questions. Some contact opportunities can also be gleaned by way of the virtual classroom. In Table 6, the existing framework for APL support is given alongside the proposed framework. The proposed framework in Table 6 highlights the more technical implementation of second-phase support, embedded support in the third and fourth phases and the constant possibility for personal email contact for interim questions.

Methodological Considerations
The study is explorative in character. The proposed support framework is based on a single case study. Therefore, we must be reserved in the generalisation of the study

results. However, the results provide a useful and informative contribution to the research on APL, which is still in its infancy. Our intention was to give support possibilities for APL in the context of distance education from the perspective of APL providers and APL candidates. The study gives many examples of types of embedded and personal support for APL purposes. Structured focus group sessions (supported by a computer system and moderated) give a lot of meaningful information. We chose to use additionally for a group of experts on online support to verify the results of the staff members and they did not find irrelevant results. Finally, the interviews with the candidates confirmed the support possibilities, and explored the institutes view on support.

In this study, we focused on the role of two main actors in APL, the staff members of the university and the candidates. A third actor, the labour field, is not taken into account in our study. This actor is in the context of APL support in many cases also an important party and can play a role in for example the translation of the APL outcome into job profiles and the use of APL in the personal development of their employers. It is important to actively involve this actor in future research.

References

Colley, H., Hodkinson, P., Malcolm, J.: Non-formal learning: mapping the conceptual terrain. Consultation report. University of Leeds Lifelong Learning Institute, Leeds (2002)
Committee flexible higher education for labour: Flexibel onderwijs voor volwassenen [Flexible Education for Adults]. Ministry of Education, Culture and Science, The Hague (2014)
Corradi, C., Evans, N., Valk, A.: Recognising Experiential Learning: Practices in European Universities. EULLearN, Estonia (2006)
Cretchley, G., Castle, J.: OBE, RPL and adult education: good bedfellows in higher education in South Africa? Int. J. Lifelong Educ. **20**, 487–501 (2001)
Day, M.: Developing benchmarks for prior learning assessment. Part 2: practitioners. Nurs. Stand. **35**, 38–44 (2001)
Denton, P., Madden, J., Roberts, M., Rowe, P.: Students' response to traditional and computer-assisted formative feedback: a comparative case study. Br. J. Educ. Technol. **39**, 486–500 (2008)
Donoghue, J., Pelletier, D., Adams, A., Duffield, C.: Recognition of prior learning as university entry criteria is successful in postgraduate nursing students. Innov. Educ. Train. Int. **39**, 54–62 (2002)
Dutch Ministry of Economic Affairs: De Fles is Half Vol. Een brede visie op de benutting van EVC [A broad vision of APL use]. The Hague (2000)
Gillies, D.: Student perspectives on videoconferencing in teacher education at a distance. Distance Educ. **29**, 107–118 (2008)
Jacklin, A., Robinson, C.: What is meant by 'support' in higher education? Towards a model of academic and welfare support. J. Res. Spec. Educ. Needs **7**, 114–123 (2007)
Joosten-ten Brinke, D.: Evaluatierapport EVC [Evaluation Report APL]. Open University of The Netherlands, Heerlen (2007)
Joosten-ten Brinke, D., Sluijsmans, D.M.A., Brand-Gruwel, S., Jochems, W.M.G.: The quality of procedures to assess and credit prior learning: implications for design. Educ. Res. Rev. **3**, 51–65 (2008)
Joosten-ten Brinke, D., Sluijsmans, D.M.A., Jochems, W.M.G.: Self-assessment in university assessment of prior learning procedures. Int. J. Lifelong Educ. **28**, 107–122 (2009)

Kawalilak, C., Wihak, W.: Adjusting the fulcrum: how prior learning is recognized and regarded in university adult education contexts. Coll. Q. **16**(1), n1 (2013)

Knowles, M.S., Holton, E.F., Swanson, R.A.: The Adult Learner: The Definitive Classic in Adult Education and Human Resource Development. Routledge, New York (2014)

Macdonald, J., McAteer, E.: New approaches to supporting students: strategies and campus based environments. J. Educ. Media **28**, 129–146 (2003)

Martens, R., Bastiaens, T., Kirschner, P.A.: New learning design in distance education: the impact on student perception and motivation (2007)

Martens, R.L., Valcke, M.M.A.: Validation of a theory about functions and effects of embedded support devices in distance learning materials. Eur. J. Psychol. Educ. **10**, 181–196 (1995)

Mason, R.: On-line learning and supporting students: new possibilities. In: Tait, A., Mills, R. (eds.) Re-thinking Learner Support in Distance Education: Change and Continuity in an International Context, pp. 90–101. RoutledgeFalmer, London (2003)

Miguel, M.C., Ornelas, J.H., Maroco, J.P.: Recognition of prior learning: the participants' perspective. Stud. Contin. Educ. **2**, 1–16 (2016). doi:10.1080/0158037X.2015.1061491

Sanseau, P.Y., Ansart, S.: New emerging issues about guidance and counselling in an accreditation of prior and experiential learning process. Int. J. High. Educ. **3**, 110 (2014)

Scheltema, L.: Rapportage project EVC-i 2002 [APL Project Report]. Stichting LOB Intechnium, Woerden (2002)

Scholten, A.M., Teuwsen, R., Mak, A.M.A.: Portfolio-ontwikkeling door buitenlandse artsen [Portfolio Development by Foreign-Trained Medical Doctors]. Nuffic, The Hague (2003)

Schuetze, H.G., Slowey, M.: Participation and exclusion: a comparative analysis of non-traditional students and lifelong learners in higher education. High. Educ. **44**, 309–327 (2002)

SER: Werk maken van scholing. [Start up schooling], The Hague (2012)

Shapiro, J.: Exploring teachers' informal learning for policy on professional development. Unpublished doctoral dissertation, RAND. (2003). http://www.rand.org/publications/RGSD/RGSD174/. Accessed 19 Apr 2004

Spencer, B., Briton, D., Gereluk, W.: Crediting adult learning. In: Proceedings of the 41st Annual Adult Education Research. Athabasca University (2000). http://www.edst.educ.ubc.ca/aerc/2000/spencerbetal1-final.PDF. Accessed 15 Nov 2004

Tait, A.: Planning student support for open and distance learning. Open Learn. **15**, 287–299 (2000)

The Calibre group of Companies. Needs assessment of a prior learning assessment and recognitionprocess for the nursing education program of Saskatchewan. Saskatchewan, Canada (2003)

Thomas, E., Broekhoven, S., Frietman, J.: EVC aan de poorten van het hoger onderwijs [APL at the gateway of higher education]. ITS, Nijmegen (2000). http://www.minocw.nl/onderwijs/evc/. Accessed 28 May 2004

Van Bruggen, J., Kalz, M., Joosten-ten Brinke, D.: Placement services for learning networks. In: Koper, R. (ed.) Learning Network Services for Professional Development. Springer, Heidelberg (2009)

Vanhoren, I.: Van herkennen naar erkennen [From identifying to recognizing]. Katholieke Universiteit Leuven, Leuven (2002)

Wheelahan, L., Miller, P., Newton, D.: Thinking about RPL: a framework for discussion. Paper presented at 11th National VET Training Research Conference, Brisbane (2002)

Wihak, C.: Prior learning assessment and recognition in Canadian universities: view from the web. Can. J. High. Educ. **37**, 95–112 (2007)

The Role of Formative Assessment in a Blended Learning Course

Sharon Klinkenberg[✉]

Department of Psychology, University of Amsterdam, Nieuwe Achtergracht 129B,
1001 NK Amsterdam, The Netherlands
S.Klinkenberg@uva.nl

Abstract. Assessing the effectiveness of a course design in higher education is an almost unfeasible task. The practical inability to conduct randomised controlled trials in a natural setting limits teachers who want to interpret their design choices to non causal evaluation. With at least a moderate amount of common sense these evaluations could be useful in gathering insight into what works and what does not work in education. In our blended learning course we wanted to assess the role of formative assessment while also taking lecture attendance into account. There are certainly many confounds for which we cannot control. We found no effect of lecture attendance but formative assessments did predict a substantial amount of course outcome. This was probably due to not including the formative performance in the final grading and our emphasis in the course that to err is to be student. The validity of the formative assessment paves the way for diagnostic use and remedial teaching.

Keywords: Formative assessment · Blended learning · Lecture attendance · Weblectures

1 Introduction

In this short paper we would like to explore the role of formative assessment in a blended learning course. Though formative assessment has a broad scope [1], we focus on online trail assessment. For the last three years a new course has been taught at the department of Psychology of the University of Amsterdam. The course Scientific and statistical reasoning combines a broad array of online and offline learning methods. We will describe the course design, scope, place in the curriculum and our attempt to assess the effectiveness of formative assessment within this course.

Blended learning can be interpreted and implemented in many ways. The interplay of many variables eventually determine a specific course outcome. Singling out one variable is therefore not easily done without a decent experimental design. However, very little time is available for lecturers in higher education to assess the methods they use. Although the generalizability of such an attempt is limited, it does provide some insight into best and worst practises. In this paper we therefore will present our results on data that is available to us while acknowledging the limited methodological robustness.

1.1 Course Design

The course Scientific and statistical reasoning is taught in the second bachelor year. Consisting of 15 ECTS (European Credit Transfer and Accumulation System) during the first semester for a body of about 400 students, it can be considered a large course according to our university standards. The course consists of four distinct cycles of four weeks each concluded with an exam in the fifth week. Figure 1 illustrates how one cycle of five weeks is designed. It also shows that students have the ability to attend a walk-in session in which specific individual issues can be addressed. Furthermore, the course combines several traditional and modern methods. There are several weekly lectures and mandatory study groups. Prior to each study group meeting an assignment needs to be uploaded to the campus electronic learning environment (ELO) and every week a digital formative assessment needs to be taken. The assessments consist of assignments in which instruction and assessment are integrated. For example, students are instructed on the subject of moderation analysis. They are presented with a case and data files which need to be used to answer the presented questions. Figure 2 illustrates such an assignment. On top of the exam students also need to write a significant argumentation essay (AE) on a specific subject.

All the different ingredients of the course are integrated in the final grading scheme by means of the following formula (1), where the final grade is determined by the mean on all four exams, the mean on the argumentation essays and deduction points.

$$\text{Grade} = .8 \times \overline{\text{Exams}} + .2 \times \overline{\text{AE}} - \text{Deduction} \qquad (1)$$

Deduction points are introduced when students do not meet the set deadlines for the formative assessment, the preparatory assignments for the study groups and study group attendance. The deduction points are presented in a graduated fashion, as can be seen in Table 1. The table shows that deduction will only take effect when multiple deadlines are not met and is created to be a reasonable

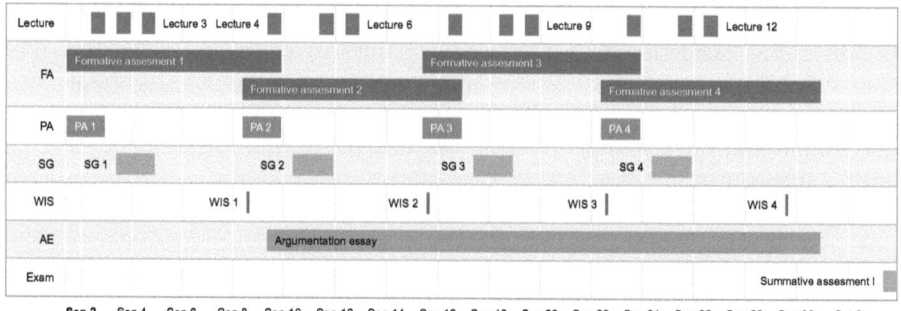

Fig. 1. Course design of cycle 1 of 4, for scientific and statistical reasoning. Showing lectures, formative assessments (FA), preparatory assignments (PA), study groups (SG), walk-in sessions (WIS), the argumentation essay (AE) and the mid term exam.

Fig. 2. Example of formative assessment question.

Table 1. Deduction points for deadlines not met. For the study groups (SG), prepatory assignments (PA) and Formative assessments (FA).

Not met	SG	PA	FA
1x	0	0	0
2x	0.25	0.1	0.1
3x	0.75	0.25	0.25
4x	1.5	0.5	0.5
5x	3	1	1
6x	Exclusion	Exclusion	Exclusion

penalty. It appeals to the phenomenon of loss aversion wherein Tversky and Kahneman [6] postulate that people tend to modify behaviour more for the avoidance of loss than for the gain of profit.

For both the formative assessment and the mandatory study groups with their prior assignment we only registered if the task was done. The quality of the work is not incorporated in the final grade. It is generally emphasized that students are allowed to make mistakes in their work but that their effort is taken into account. In recent years we encountered massive harvesting of our formative assessment items due to the fact that performance marginally determined the final result. Even our good students who did spend time on the assignments resorted to looking up the correct answer before submitting the final answer. The results on the formative assessment did not show any predictive validity with the course exams and therefore could not be used as a diagnostic instrument for assessing individual problems in certain subdomains. We emphasize that making errors is part of the learning process and that with sufficient effort the course objectives can certainly be met. With this approach we aim to appeal to a growth mindset [3] that states that people who believe ability is not rigid have less problems obtaining new knowledge.

Apart from the above described course design, students also have access to the weblectures from our campus video lecture capture system. These weblectures are available directly after the lecture has been given.

2 Methods

In our analysis we aim to identify the role of formative assessment in the scope of the above described blended learning course. There are a lot of confounds and we can only present results on the limited variables we have access to.

The variables that determine the final course grade e.g. exams and argumentation essays are used as validation criteria. The deduction points are not taken into account as they indirectly determine the correlation between course outcome and predictors and therefore violate conditional independence. This final course grade is a non unidimensional indication of course performance. At least two kinds of skills can be theoretically extracted, namely the ability to perform and interpret statistical analysis and general scientific reasoning ability. For the statistics part SPSS is taught and R [5] is used for manual calculations. Furthermore, both the exam items as well as items used in the weekly assignment are tagged on subcategories. We use the scores on these subcategories to inform students on their performance.

2.1 Exam Results

There are four midterm exams which together determine 80% of the final grade. An exam consists of a statistics part and a critical thinking part and are administered digitally. An entire exam is constructed of multiple choice questions, fill in the blank questions and small essay questions. Furthermore, students are required to download data files and run analysis in SPSS and interpret the output from said analysis. Figure 3 shows an exam item as used. The results of the exams are subjected to general quality control analysis and corrected when necessary. Finally, items are tagged to categories and these sub scores are made available to students.

2.2 Mandatory Tasks

The study group attendance, preparatory assignments and formative assessments are all mandatory and result in course exclusion if deadlines are not met. They are scored as 0 or 1 and are incorporated as deduction points in the final grade as shown in Table 1. The inclusion of these binary results will not add any

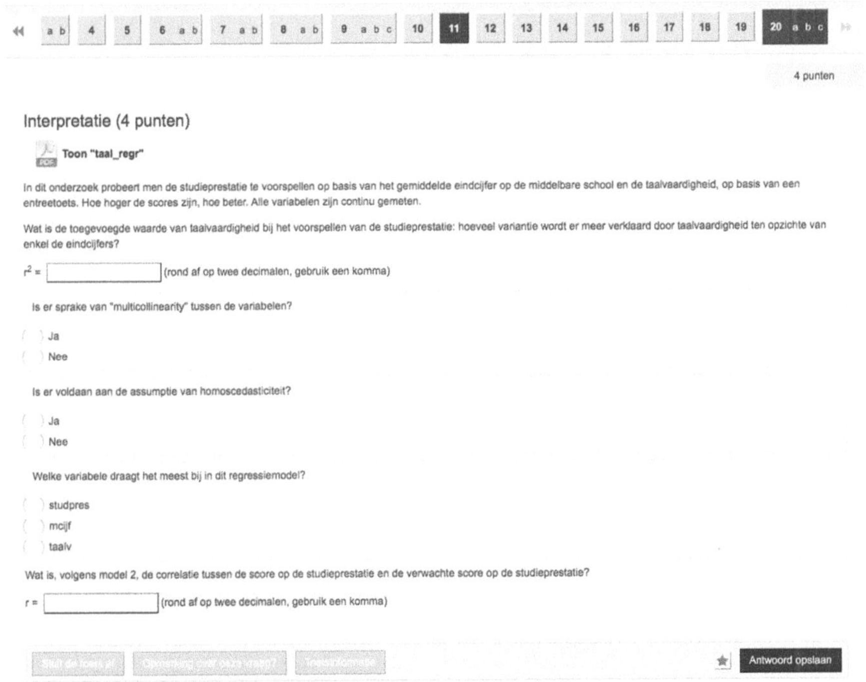

Fig. 3. Example of summative assessment item.

predictive value in course performance due to the fact that all students comply and only some are excluded from the course. We therefore will not add these binary variables to our model.

2.3 Formative Assessment

Apart from the binary compliance score the formative assessments also produce a score indicating performance about the whole assessment and also on all subcategories. These results will be taken into account. For the entire course there are 14 of these assessments and a general formative assessment indicator is created by extracting a factor using a principal component analysis (PCA). We used this method to arrive at an optimally fitting aggregated assessment facter with minimal bias.

2.4 Lecture Attendance

Though lecture attendance was not compulsory we did register course attendance in the first and final cycle of the course. This was done to get some indication of attendance drop during the course like we had experienced in the previous year. Students were asked to submit their student id at the beginning of the lecture.

We scored the attendance for every student as zero or one. Again we used the PCA method to arrive at a single factor for attendance.

2.5 Weblectures

Only the anonymous data on weblecture viewing behaviour was available to us. We therefore were not able to relate course performance to this variable. Our department feared that students whom did not attend campus lectures would procrastinate and would under achieve in this course. We analyse the anonymous results to get an insight into viewing behaviour.

In our analysis we will look at the formative assessment results to the lecture attendance, furthermore we will fine-grain the predictive validity of the formative assessment and finally we will eyeball the weblecture viewing behaviour.

3 Analysis

Our analysis is based on the $N = 427$ (70% female) students that attended the course. The average age for males and females was about 21 (sd males 2.14, sd females 2.7). Where the youngest student was 18 and the oldest 47. By far most students were 20 years of age.

We will first take a general look at the variables we can use to predict course performance which are lecture attendance and formative assessment results. There were 14 formative assessments during the course and we registered the attendance on 15 lectures, of which all lectures given in cycle 1 and all lectures in cycle 4. To reduce the amount of data we used a principal component analysis to extract one factor for course attendance and one for the formative assessment. Both factors were used in a regression analysis to predict the final grade corrected for the deduction points. This model explained 29% $F(2, 271) = 55.2, p < 0.001$ of the variance but looking at the beta coefficients revealed lecture attendance did not contribute to the model $b = -0.006$ $t(271) = -0.31, p > .05$. The coefficient for the formative assessment scores was $b = 0.22$ $t(271) = 10.49, p < .001$ (Eq. 2). For this regression analysis 150 observations were dropped due to missing values. This resulted from the PCA not being able to retrieve a factor score due to lack of variance or subjects not having a score on one of the three variables used.

$$\widehat{FG} = 6.55 - .006\,LA + .22\,FA \qquad (2)$$

It is fair to assume that lecture attendance in cycle one and four could not contribute to the total course performance because attendance was not measured in the second and third cycle. Therefore, we also analysed the predictive value on both predictors solely on the first and final exam. Here we used all course attendance results for the first cycle to predict the grade at the first midterm exam. Again attendance explained no variance $R^2 = .03$ while the four formative assessment results explained 24% $F(380) = 31.7, p < .001$. Cycle four showed that both attendance and formative assessment did not explain the grade for

Table 2. Regression R^2 for predicting final grades (FG) and mid term grades (MT) one to four with lecture attendance (LA) and/or formatieve assessment (FA) in the model.

Model	MT1	MT2	MT3	MT4	FG
LA + FA					0.28
LA	0.03			0.05	
FA	0.25	0.23	0.10	0.05	

the fourth midterm exam. Both models explained less than 5% of the variance. Though both were significant the effect is negligible.

For cycle two and three no lecture attendance data was available. We therefore only analyzed the results of the formative assessments. These respectively explained 23% $F(4, 368) = 27.74, p < .001$ and 10% $F(4, 374) = 10.89, p < .001$ of the variance. Table 2 summarizes the explained variance per model. Reliability of the four exams ranged from .53 to .68 on Cronbachs alpha.

3.1 Weblectures

It can be assumed that lecture attendance did not contribute to course outcome due to the fact that many students resorted to the weblectures. A combination of weblecture and attendance data would probably show that at least viewing the lectures would contribute to performance. Due to the privacy policy of the university we were not able to obtain the personalised results, therefore this combined insight could not be attained. As stated, the drop in lecture attendance had the department worried about course outcome due to procrastination. If this were the case we would expect the viewing frequencies to spike in the week leading to the exams. Figure 4 clearly shows the viewing frequency to be centred within the week of the lecture. The graph displays the frequency per day for the entire duration of the course for all given lectures. We only see a slight dent in the week leading up to the exam. We can therefore conclude that procrastination is no issue within this blended learning course.

4 Discussion

The results show that formative assessment plays a notable role in predicting course performance, though not all cycles contributed equally well. In particular the final cycle showed poor performance. We attribute this to the limited amount of formative assessments in this final cycle. Compared to the rest of the course, only two assessments were administered. Another component here is that the final exam also consisted of items containing topics on the previous cycles. The amount of lectures was also the lowest in this final cycle, though here the argument of exam content does not hold.

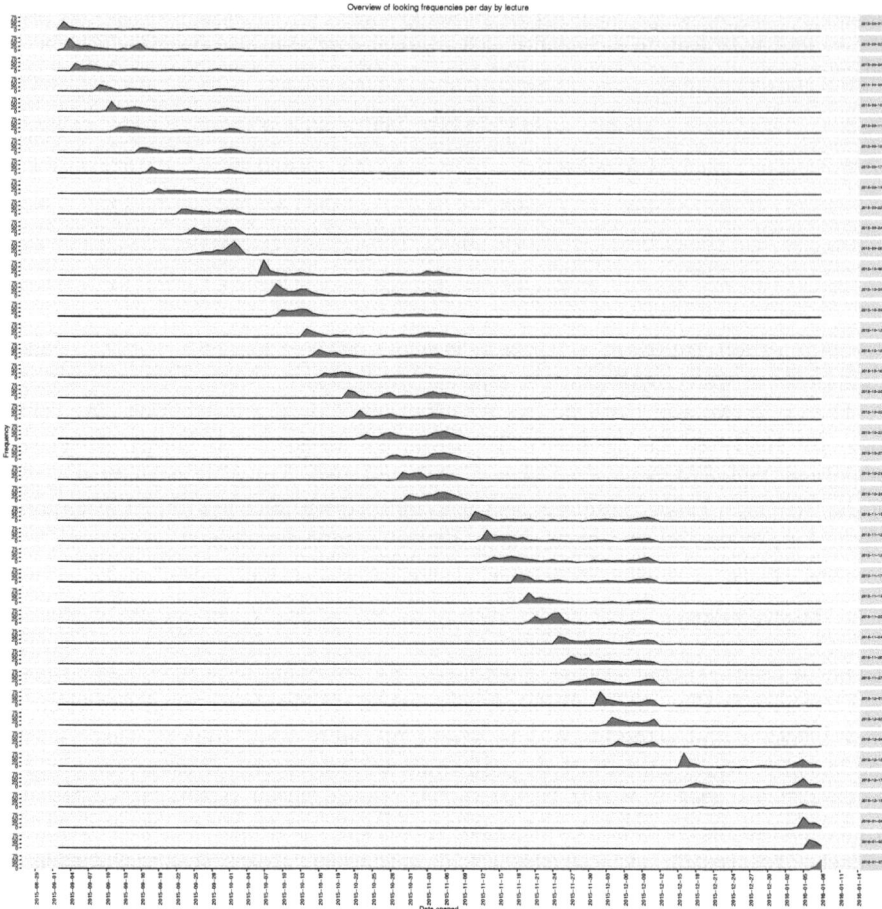

Fig. 4. Weblecture viewing frequencies for all lectures y-axis and course duration x-axis.

As stated earlier, the lack of any predictive value for lecture attendance could be because the weblecture data was not taken into account. It would be reasonable to at least assume that viewing the lecture content would influence course outcome.

That assumption does lead us to the question if attendance equals retention of content. Research on retention by Freeman et al. [4] at least shows that lectures without any form of active learning are not that effective. Our panel group discussions with students also indicate that not all students benefit equally well from lectures. Some find it hard to concentrate or to keep track. Others mention being distracted by other students. On the other hand students report the weblectures to create some differentiation. Our top students mention not attending the lectures and opting to watch the weblectures at double speed, while our struggling students use the weblectures to revisit the content more often. It is therefore not unreasonable to expect weblecture views to be a better predictor of course outcome than lecture attendance. It should be strongly noted that this expectation only holds in the context of a blended learning setup in which pacing is highly controlled. Though research by Bos and Groeneveld [2] on the effects of video lectures show no difference compared to lecture attendance.

5 Conclusion

The results of this analysis show that formative assessment has a decent amount of predictive value when it comes to course outcome. Not incorporating the performance on the formative assessment in the calculation of the final grade, and emphasizing that errors are allowed seem to have worked to restore predictive validity. It can also be concluded that a sufficient amount of formative assessment needs to be administered and that the exams need to be properly aligned with the formative assessment. In the coming academic year we will start using the formative assessments as a diagnostic tool to facilitate remediation.

References

1. Black, P., Wiliam, D.: Developing the theory of formative assessment. Educ. Assess. Eval. Account. **21**(1), 5–31 (2009)
2. Bos, N., Groeneveld, C., Bruggen van, J Brand-Gruwel, S.: The use of recorded lectures in education and the impact on lecture attendance and exam performance. Br. J. Educ. Technol. (2015)
3. Dweck, C.S.: Self-theories: Their Role in Motivation Personality and Development. Psychology Press, Philadelphia (2000)
4. Freeman, S., Eddy, S.L., Mcdonough, E., Mcdonough, M., Smith, M.K., Okoroafor, N., Jordt, H., Wenderoth, M.: Active learning increases student performance in science, engineering, and mathematics. In: Proceedings of the National Academy of Sciences of the United States of America (2014)
5. R. Core Team: R: a language and environment for statistical computing (2016)
6. Tversky, A., Kahneman, D.: The framing of decisions and the psychology of choice. Science **211**, 453–458 (1981)

Self- and Automated Assessment in Programming MOOCs

Marina Lepp[✉], Piret Luik, Tauno Palts, Kaspar Papli, Reelika Suviste, Merilin Säde, Kaspar Hollo, Vello Vaherpuu, and Eno Tõnisson

Institute of Computer Science, University of Tartu, Tartu, Estonia
{marina.lepp,piret.luik,tauno.palts,kaspar.papli,
reelika.suviste,merilin.sade,kasparho,vellov,eno.tonisson}@ut.ee

Abstract. This paper addresses two MOOCs in Estonian about programming where different kinds of assessment were used. We have used two kinds of automated assessment: quizzes in Moodle and programming exercises with automated feedback provided by Moodle plug-in VPL. We also used two kinds of self-assessment: (1) self-assessment questions with feedback and explanations for every answer and (2) so-called "troubleshooters" for every programming exercise, which contain answers to the questions that can arise during the solution of a given exercise. This paper describes our experience in the creation of quizzes, programming exercises, and tests for automated feedback, self-assessment questions, and troubleshooters. The paper discusses the problems and questions that arose during this process and presents learners' opinions about self- and automated assessment. The paper concludes with a discussion of the impact of self- and automated assessment in MOOCs, describes the work of MOOC organizers and the behaviour of learners in MOOCs.

Keywords: Self-assessment · Automated assessment · MOOC · Programming

1 Introduction

Computer programming is a current topic of interest in Estonia as well as in other countries, and massive open online courses (MOOCs) offer a good opportunity for everyone to get acquainted with this subject [17]. This paper addresses two MOOCs on programming (4 week course "About Programming"; 8 week course "Introduction to Programming"), provided in Estonian language and intended primarily for adults. The MOOCs turned out to be quite popular and have been organized several times since 2015.

Traditionally there were two main kinds of MOOCs: cMOOC (courses for connecting people) and xMOOC (courses with structured content) [1, 15]. However, there is a move away from the cMOOC/xMOOC division to recognition of the multiplicity of MOOC designs, purposes, topics and teaching styles [1]. Our courses are not very traditional xMOOCs, as in addition to a forum the learners have the opportunity to ask the helpdesk provided by the organizers about their programs, problems, etc. Forums are used widely in MOOCs for getting help and advice [3], but previous MOOCs that used a helpdesk were rated extremely positive [18]. In our case this opportunity resulted in a high completion rate (50–70% in different courses at different times), which is unusual for MOOCs. Research has shown that the average completion rate for MOOCs is approximately 15% [9] and dropout rate is one of the challenges of MOOCs [15]. However, the cost of providing this opportunity is very high for the organizers.

We hoped that different kinds of assessment would reduce the number of learner emails with questions to organizers, while maintaining a high completion rate. Assessment is one of the emerging key themes of MOOC [4] and may be a key driver for learning [11]. Different types of assessment are used in different MOOCs [11], for example automated assessment [19, 12], peer assessment [10] and self-assessment [20]. Learners favour a combination of assessment types, as each of them serves a different purpose for their learning [11].

Automated assessment is one of the key issues in MOOCs, because it is very time-consuming to check exercise solutions from a large number of students by hand [15]. It has been observed that learners recognise the benefits in peer assessment but they prefer automated assessment [11] and especially programming MOOC can benefit from automated assessment tools [17]. We have used two different kinds of automated assessment. One of them included quizzes with multiple-choice and short-answer questions in the Moodle e-learning environment [5]. Previous research has found that weekly quizzes significantly explained differences in students' final exam scores [1]. The other automated assessment type included programming exercises with automated feedback provided by Moodle's Virtual Programming Lab plug-in [14], which allows instructors to formulate programming assignments and specify tests for giving automatic feedback on solutions. Immediate results and feedback to the students provided by automated assessment is extremely valuable when learning to program [6], but providing high quality automated assessment of programming assignments can be very challenging and demands increased effort from the instructor [12].

Self-assessment is also very important in MOOCs, and especially in our case, as we received a lot of letters with questions. It has been suggested that self-assessment should be used as assessment *for* learning instead of assessment *of* learning [1]. Our course material contains self-assessment questions with feedback and explanations for every (right and wrong) answer, so using these effectively could reduce the amount of individual questions. The second kind of self-assessment was provided only in the courses given in 2016. We provided so-called "troubleshooters" for every programming exercise. Troubleshooters contain answers and clues to the questions, which can arise when solving a particular exercise. A similar kind of self-assessment (exercises with built-in scaffolding messages inside the programming environment) has been considered rewarding in a programming MOOC [17].

This paper describes our experience in the creation of quizzes, programming exercises, and tests for automated feedback, self-assessment questions, and troubleshooters. The paper gives an overview of the problems and questions that arose during this process and also presents learners' opinions about self- and automated assessment. We requested feedback with the help of a survey several times during every course to investigate participants' opinions about the course, learning styles, learning tools, and different types of assessment used in MOOCs.

There are 6 specific research questions of interest. How much do learners use self-assessment questions and troubleshooters in their own opinion? How helpful is self-assessment for learners in their opinion? Which learners report using self-assessment and who found these tools more helpful? How does the use of self-assessment correlate with performance on automatically assessed tasks? Can self-assessment reduce the amount of letters/questions to organizers? How do different kinds of assessment affect the work of MOOC organizers?

The paper concludes with a discussion of the impact of self- and automated assessment in MOOCs, the work of MOOC organizers, and the behaviour of learners in MOOCs.

2 Research Methods

2.1 Background of Courses

This article describes two MOOCs that have been organized by the Institute of Computer Science at the University of Tartu. The institute has been developing MOOCs since the winter of 2015. These two MOOCs are offered in Estonian language. The courses are not strictly traditional MOOCs, as in addition to a forum the learners have an opportunity to ask the helpdesk provided by the organizers about their programs, problems, etc. The courses are developed, maintained and run by a team of 14 faculty members, teachers and assistants.

The first MOOC is named "About Programming" (in Estonian *Programmeerimisest maalähedaselt*); it lasts for 4 weeks and the amount of expected work is 26 h. The course is designed primarily for adults and the targets people who have no previous experience with programming. "About Programming" aims to introduce basic programming concepts and structures. This MOOC also introduces IT related subjects and examples like self-driving cars, language technology etc. So far this MOOC has been organized 4 times. The first course was a trial and was aimed at developing and testing the materials and exercises, which was the main reason for the somewhat low participation rate. The second time the number of learners was also limited, due to the fact that the organizers were not confident in how well the process will turn out (especially in terms of managing letters from learners). After that the number of learners was not limited. The number of registered learners and the number of learners, who actually start the course (by submitting at least one exercise or test), the corresponding number of learners completing the course and the completion rates are shown in Table 1.

Table 1. Number of learners, the corresponding number of completing learners and completion rates in the course "About Programming"

Time	Registered	Started	Completed	Completion rate among registered	Completion rate among started
Winter 2015	32	30	21	66%	70%
Spring 2015	638	547	411	64%	75%
Autumn 2015	1,534	1,270	1,010	66%	79%
Spring 2016	1,430	1,224	885	62%	72%

The second MOOC is named "Introduction to Programming" (in Estonian *Programmeerimise alused*); it lasts for 8 weeks and the amount of expected work is 78 h. This course is also intended for learners who have little or no experience with programming. Previous completion of the course "About programming" is useful, but is not absolutely necessary if a learner is highly motivated and eager to learn. "Introduction to Programming" gives an overview of some programming techniques (providing a much deeper insight into the topics of the course "About programming") and teaches the basics of algorithmic thinking. So far this MOOC has been organized twice since the winter of 2016. The first time was a trial and the number of learners was limited. The second time the number of learners was not limited. The numbers of this MOOC are shown in Table 2.

Table 2. Number of learners, the corresponding number of completing learners and completion rates in the course "Introduction to Programming"

Time	Registered	Started	Completed	Completion rate among registered	Completion rate among started
Winter 2016	295	278	145	49%	52%
Spring 2016	1,770	1,523	970	55%	64%

2.2 Assessment

The material of the courses consists of programming material with video clips and self-assessment questions and material of general interest. During every week learners should do programming exercises with automated feedback (1–2 exercises in "About Programming", 2–4 exercises in "Introduction to Programming") and one quiz. Every exercise was provided with a so-called "troubleshooter" with answers to the questions, which can arise when solving a particular exercise. While learners can choose whether they want to use the troubleshooters and to answer self-assessment questions, the programming exercises and quizzes are compulsory in order to pass the course.

Automated Assessment. Automated assessment plays an essential role in MOOCs, because it is very time-consuming to check the exercise solutions from a large number of students by hand. We have used two kinds of automated assessment. One included quizzes with multiple-choice and short-answer questions in the Moodle environment. The other included programming exercises with automated feedback provided by Moodle's Virtual Programming Lab plug-in, which allows instructors to formulate programming assignments and specify tests for giving automatic feedback on solutions.

Weekly Quizzes. The main purpose of the quizzes is to make sure that the student has read and understood the general learning material of the course (programming and additional information). Each weekly quiz has 10 questions about MOOC course learning material. The quiz is passed if 90% (9 questions out of 10) of the answers are correct. The quizzes mainly comprise multiple-choice questions with examples of short programs to confirm that the learner can understand and analyse programs. The quizzes also include open-ended questions, requiring short answers (e.g., a number).

Depending on the answer, the student receives positive or negative feedback. The question behaviour that is used in our quizzes is "interactive with multiple tries" [7]; it means that feedback is received directly after submission and the number of submissions is not limited. After submission, if the answer is incorrect, the correct answer is not given, but a negative feedback with a helpful hint for the right answer is given. Unlimited quiz submissions per student and hints enable students to learn from feedback and correct their answers. However, the downside is that it allows learners simply to try every answer to find the correct solution without really thinking about the quiz responses. This is the main difficulty and problem with quizzes in MOOC. It has been suggested that some learners in online courses are "solvers" in terms of their engagement style [2] and they primarily hand in assignments for a grade, viewing few if any of the lectures. This group can also include a subcategory of "clickers", who do not know the right answer and just click for it, but further research is needed to confirm that. Another problematic question concerning quizzes is how to create good questions to confirm that students have understood the course material.

Programming Exercises with Automated Feedback. Weekly programming exercises require creation of a computer program as the outcome. The exercise specifies the input and output data of the program and sometimes also some additional requirements or constraints that the program must comply with. Input and output of the program is typically defined as a sequence of strings that must be read from or written to standard input-output streams or text files. Therefore the correctness of the resulting program is defined by whether or not it outputs the expected data given some input data. This is very convenient since it allows us to automatically feed predefined input data in the submitted programs and check if the output is correct for that particular exercise. If a test case fails, automatic feedback can be given about what was expected of the program and how it actually performed.

The number of submissions is not limited to allow students to learn from feedback and correct their solution. We use Virtual Programming Lab [14] for this kind of automated assessment. It allows us to define test cases for each exercise that contains the input and expected output of the program and feedback messages for different situations. During the design and creation of automatic assessment tests we faced three critical requirements: the tests must accurately verify the correctness of the solution, they should give constructive feedback and they should not impose strict constraints on the output formatting or program architecture. Each of these requirements is further discussed in detail.

Firstly, the automatic tests must be very accurate since the submissions are not reviewed by humans. This means that any submission that passes the tests must be correct in terms of the exercise and any submission that fails to pass at least one test must be incorrect. If submissions were reviewed by staff then the tests could remain imperfect such that a solution could be accepted manually even if it does not pass all tests, but no such exception can be made in the case of automatically assessed MOOC exercises with thousands of submissions.

Secondly, the tests must provide constructive comments because that is the main source of feedback for the students whose submission is not accepted. Assuming that the data used for testing a solution is public, basic feedback such as "Input: aaa. Expected program to output xxx but got yyy instead" can always be given. Sometimes more advanced feedback can also be very useful, for example: "Function aaa was defined but it should also be called inside the main program."; "Output of the program was correct but variable bbb was not used. This variable should be used in the output message."; "This exercise does not require a user input so the 'input()' function should not be used." It would be even better to automatically derive minimal corrections to student's incorrect solutions providing them with a measure of exactly how incorrect a given solution was [16], but VPL does not enable us to do this.

Thirdly, the technical constraints on the solution must be as gentle as possible. Testing program output means determining if the output contains the expected data and does not contain any unexpected data. This can be done easily if the output format of the data is clearly defined. However, following strict rules for the output format can often be counterintuitive for students with no background in exact sciences. For example, a student might choose to use two empty lines between the first and second output messages instead of one or even choose a synonym for some word. Based on student feedback we have noticed that these types of mistakes can be very discouraging since the submitter's correct solutions are claimed to be faulty. Therefore, we have attempted to design tests such that these technical constraints are kept to a minimum.

Catering for all three requirements simultaneously is a challenge. Designing of tests very often involves trade-offs between these requirements. For example, accurate constructive feedback and even suggestions could be given if the architecture of the solution was fixed. However, there are many different solutions to a given problem, which means that defining a 'correct' architecture is not an option. All possible correct solutions must be considered. Extracting these pieces of logic that are invariant over all correct solutions has been one of the most difficult challenges for us when designing both automatically graded exercises and tests.

Self-assessment. Self-assessment is also very important in MOOCs, and especially in our case, as we received a lot of letters with questions. The course material contains self-assessment questions with feedback and explanations for every (right and wrong) answer. The second kind of self-assessment was provided only in the courses offered in 2016. We provided so-called "troubleshooters" for every programming exercise. Troubleshooters contain answers and clues for the questions, which can arise when solving a particular exercise.

Self-assessment Questions. Programming materials of both courses contain self-assessment questions. Self-assessment questions are designed to help learners self-evaluate the essential programming points of the material. Feedback is provided for every answer. Notably, the feedback does not only specify whether an answer was wrong or right, but it also explains for every answer why this answer is wrong or right (Fig. 1). Using it effectively can reduce the amount of individual questions. Weekly videos and letters encouraged students to study feedback on every answer.

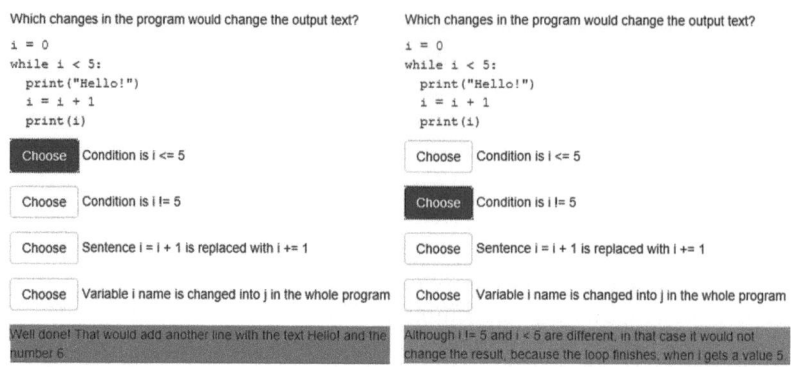

Fig. 1. Self-assessment question with explanation when right (left part) or wrong (right part) answer is chosen

Different teaching methods are used in composing the material with self-assessment questions. The learning material is structured in different ways so that sometimes there is material with a new topic and then self-assessment questions and sometimes the self-assessment questions are presented at first and then the material explaining a new topic. The teaching method "learning by mistake" is used in some self-assessment questions, as it is a very powerful method in teaching programming [8]. The questions have to be interesting; every (right or wrong) answer has to have a didactic value. Composing such questions is not an easy process and is certainly a challenge for the organizers.

Our own environment was created for making self-assessment questions. The environment produces ordinary HTML code, which is convenient for copying and pasting into materials. There are different types of questions that are created – questions with one right answer, questions with several right answers, questions in which the right order of answers should be marked. The use of self-assessment questions is not logged and unfortunately the organizers do not have the exact picture of the use of self-assessment questions by students. The participants only reported their use and opinion about self-assessment questions in questionnaires.

Troubleshooters. A helpdesk was created to offer students the opportunity to use help 24/7. A rule was established that each question had to be answered by the organizers in 8 h, which could have contributed to a high completion rate of the course (around 60%). However, such an approach is very expensive as it involves many staff members and university students proficient enough to be engaged in the feedback process.

Another positive aspect of the helpdesk was that the letters with the questions were collected and the questions of learners were analysed. The questions were categorized based on existing classifications (e.g., [13]) and opinions of the course organizers. As a result of the categorization, an environment called *Murelahendaja* (Troubleshooter) was created in 2016 by Vello Vaherpuu to help the participants with the questions that may arise during the course.

Murelahendaja is an environment, which allows finding solutions to the problems in an option tree format. It can be used for various problems and was used with problems from registering to the course to finding a solution to specific error messages. As a result every programming task was supported by the troubleshooting system (Fig. 2, left part). The troubleshooters and the environment were tested throughout the course in order to give as many sensible answers (Fig. 2, right part) to the questions as possible. It must be pointed out that in order to have an educational effect, the troubleshooters cannot give straight solutions to the tasks, but should help with systematic hints.

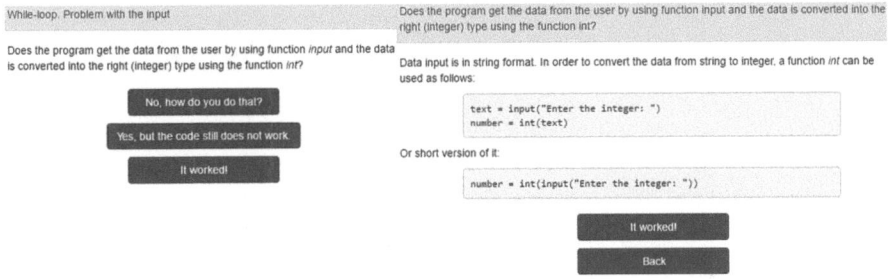

Fig. 2. Part of a troubleshooter with a question (on the left) and with an answer (on the right)

2.3 Questionnaires

Participants' opinions about the course, learning styles, learning tools, and different types of assessment used in MOOCs were surveyed using questionnaires. We collected feedback data at the end of the course from 792 participants (89.5% of completing learners), who completed the course in the spring of 2016. From the participants 342 (43.2%) were male and 450 (56.8%) female and 790 (99.7%) were from Estonia. The average age of the participants was 35.8 years (SD = 10.9) ranging from 12 to 81.

The participants filled in the questionnaire evaluating how difficult they perceived the last three exercises and the last weekly quiz (we did not ask about any previous exercises, because the students might not remember the difficulty of those at the end of the course), how much they used self-assessment questions and troubleshooters and how helpful these tools were for them. The Moodle learning analytics of each participant, indicating the attempts to submit tasks and the points for tasks, was matched to the answers from the questionnaire and to the background data from the pre-questionnaire (how much they think that mathematics is something for them and how much they like programming).

2.4 Analysis Strategy

Statistical analyses were carried out as follows. First, descriptive statistics on the participants' background information and opinion on using different assessment tools was investigated. Next, Spearman correlation coefficients were calculated to investigate the relationship between self- and automated assessment in MOOCs, but also between other assessment tools. The helpfulness of self-assessment for learners was also investigated using the Spearman correlation coefficient.

The analyses were carried out using the statistical package SPSS version 23.0.

3 Results

The results (see also Fig. 3) indicated that most participants used self-assessment questions (96.1% stated that they used self-assessment), but not all read the explanations on self-assessment (62.8% agreed with the respective statement). Troubleshooters were used by 55.8% of those who answered the questionnaire. Obviously, all students do not need troubleshooters.

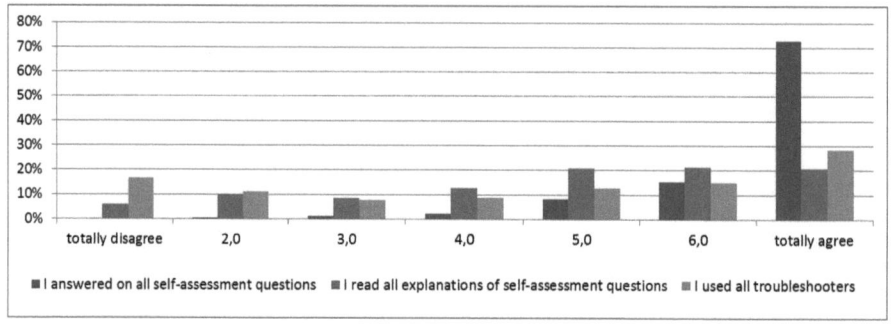

Fig. 3. Participants' responses on using self-assessment questions and troubleshooters

In order to describe the perceived helpfulness of these tools, we only used data from the participants who reported (by selecting response 4 or higher) that they used these tools. 94.5% of the participants, who had used self-assessment, agreed that this tool is helpful and almost the same percentage (95.5%) of those, who had read self-assessment explanations, found self-assessment helpful. Troubleshooters were found helpful by 84.7% of the respondents who had used them.

The Spearman correlation coefficients were .488 ($p < .01$) between confirming the use of self-assessment and finding self-assessment helpful; .162 ($p < .01$) between confirming the reading of explanations of self-assessment and finding self-assessment helpful; and .746 ($p < .01$) between confirming the use of troubleshooters and finding the troubleshooters helpful. The results indicate that the reading of self-assessment explanations was modestly correlated with finding self-assessment helpful.

In order to identify, which participants used these tools as reported by themselves and who found these tools more helpful, we calculated Spearman correlation coefficients between these reported answers and the participants' background data (see Table 3). The results reveal that most of the correlations are quite weak. Only the reported pleasantness of programming was correlated to reading explanations of self-assessment and using troubleshooters ($\rho = .261$) – it indicates that if you like something you try to find any help to achieve the desired goal. An interesting result was that those participants, who indicated that mathematics is something for them, used less troubleshooters but reported a higher use of self-assessment.

Table 3. Spearman correlation coefficients between participants background data and their responses on the use of self-assessment questions and troubleshooters

	Reported use of self-assessment questions	Reported reading of explanations on self-assessment questions	Reported helpfulness of self-assessment questions	Reported use of troubleshooters	Reported helpfulness of troubleshooters
Age	.026	.088*	.099**	−.004	−.022
Agreed that mathematics is something for me	.092*	.012	.042	−.128**	−.059
Agreed that programming is pleasant	.003	.120**	.058	.261**	.187**

* Statistically significant on .05 level
** Statistically significant on .01 level

The Spearman correlation coefficients between perceived difficulty of the last exercises, the last weekly quiz and the reported use of self-assessment and troubleshooters are presented in Table 4.

Table 4. Spearman correlation coefficients between the perceived difficulty of the last exercises/final quiz and their reported use of self-assessment questions and troubleshooters

	Reported use of self-assessment questions	Reported reading of explanations on self-assessment questions	Reported helpfulness of self-assessment questions	Reported use of troubleshooters	Reported helpfulness of troubleshooters
Perceived difficulty of last exercises	−.026	.103**	.033	.348**	.262**
Perceived difficulty of the last weekly quiz	−.057	.115**	−0.15	.174**	.113**

** Statistically significant on .01 level

We calculated how many total attempts each participant made in order to submit correct solutions of exercises and the correct answers to weekly quizzes, and how many points they earned from weekly quizzes, and then calculated Spearman correlation coefficients between these variables and the reported use of self-assessment questions and troubleshooters (see Table 5).

We asked in the questionnaire from the participants, how much they agree, that they focused on solving exercises and weekly quizzes. These answers were negatively correlated with the reported use of self-assessment (rho = −.084; p < .05), but positively correlated with the reported on use of troubleshooters (rho = .107; p < .01).

Table 5. Spearman correlation coefficients between participants' learning analytics and their reported use of self-assessment questions and troubleshooters

	Reported use of self-assessment questions	Reported reading of explanations on self-assessment questions	Reported helpfulness of self-assessment questions	Reported use of troubleshooters	Reported helpfulness of troubleshooters
Number of attempts to submit correct solutions of exercises	−.118**	.145**	−.078*	.300**	.247**
Number of attempts to submit weekly quiz (at least 90% of right answers)	−.011	.053	.015	.146**	.113**
Total points in weekly quizzes	.104**	.000	.072*	−.223**	−.183**

* Statistically significant on .05 level
** Statistically significant on .01

4 Discussion and Conclusion

The first research question was about how much learners use self-assessment. As self-assessment is not compulsory for passing the course, learners can actually decide whether they want to use self-assessment or not. The results indicated that almost all participants used self-assessment questions and many of them (but not all) read the explanations on self-assessment answers. The importance of assessment was highlighted also by previous studies (e.g., [4, 11]). Roughly half of the participants used troubleshooters. It is obvious that not all of them need troubleshooters, but troubleshooters as one possibility to replace helpdesk could influence the attitude towards the MOOC [18] and could be one reason why in our MOOC the dropout rate was lower than in most MOOCs [9, 15]. The second research question concerns the helpfulness of self-assessment for learners in their opinion. The option of self-assessment received very positive feedback from the participants in their answers to the surveys. These participants, who used self-assessment (both self-assessment questions and troubleshooters), indicated that these tools were helpful for them. Combining self-assessment with other assessment tools can serve different purposes in learning [11]. In our case this tool was created for learning as was suggested by Admiraal et al. [1] and therefore could be helpful for learners. We also wanted to study the characteristics of the participants (based on their background data), who reported using self-assessment and who found these tools to be more helpful. The results indicate that only the reported pleasantness of programming is correlated with reading explanations on self-assessment and using troubleshooters – if you like something, you try to find any help to achieve the desired goal. However, the fact that learners' behaviours and intentions are very diverse must be taken into account [1]. An interesting result was that those participants, who agreed with the statement that mathematics is something for them, used less troubleshooters but more self-assessment questions.

The next findings pertain to the correlation between the use of self-assessment and performance on automatically assessed tasks. The results indicated that the participants, who made more attempts to submit the right solutions of exercises and correct answers to weekly quizzes, used more troubleshooters and found troubleshooters to be more helpful, but they used less self-assessment questions and found them less helpful. The results of previous studies [1] have indicated that the scores on the quizzes showed low to moderate positive correlations with self-assessment and also the correlations between the number of different assessment attempts and the final-exam grade were positive, but moderate to low, which differs from the result of our study. Therefore, it is possible that we have to direct participants to self-assessment questions after a certain number of unsuccessful tries to submit an exercise solution or a weekly quiz, because not only troubleshooters but also self-assessment questions with answers can help participants.

The study showed that the two tools of self-assessment (self-assessment questions and troubleshooters) complement each other and can be suitable for different participants. The participants, who perceived exercises more difficult, used more troubleshooters and found the troubleshooters to be more helpful, but used less self-assessment questions and found self-assessment questions to be less helpful. The participants, who earned more points on weekly quizzes, used more self-assessment and found this to be more helpful, as was also found in a previous study [1], but they used less troubleshooters and found them less helpful. As has been suggested in previous studies [2], learners differ in the ways they engage with online courses. Some participants use more self-assessment and acquire the needed knowledge through that without needing the troubleshooters, while other participants (called "solvers" in [2]) do not want to use self-assessment and focus on solving exercises and weekly quizzes, using troubleshooters if they encounter problems. This finding is supported by the next result of our study. The participants' reported level of agreement with the statement that they focused on solving exercises and weekly quizzes, was negatively correlated with the reported use of self-assessment, but positively correlated with the reported use of troubleshooters.

The next research questions addressed the work of the organizers. We found that self-assessment can reduce the amount of letters/questions to the organizers. As a result of using troubleshooters the number of letters to the organizers decreased from 1,250 during a 4-week course (in the autumn of 2015) to 750 (in the spring of 2016), representing a decrease from 0.8 to 0.5 letters per participant. As troubleshooters reduced the number of letters with questions to the organizers, the course organizers believe that this kind of self-assessment is very useful for both learners and organizers. We also found that, while the composition of quizzes was quite easy for the organizers and answering them was easy for the learners, the programming exercises were a challenge to both the organizers and the learners. We agree with other researchers that providing high quality automated assessment of programming assignments is very challenging and demands increased effort from the organizers [12]. However, automated assessment makes it possible to organize programming MOOCs and programming MOOCs certainly benefit from the application of automated assessment [12].

This study has some limitations that have to be taken into account when drawing generalisations from the findings. First, a clear limitation of this study is the limited number of MOOCs examined. Further research should investigate more MOOCs. Second, the use of assessment tools was partially self-reported. Although the participants' ratings on the use of troubleshooters and self-assessment provided important information, they did not describe the actual use of these assessment tools. Finally, the results and findings of this study are based on a post-questionnaire, which was filled out only by those participants, who successfully passed the course. The next study should also examine the opinions of the participants, who dropped out the course.

In conclusion, the results of the study indicate that self-assessment questions and explanations on self-assessment answers are of importance. Different learners prefer different assessment tools, which they find helpful for completing the course. The study showed that the use of self-assessment was related to the performance on automatically assessed tasks. Our findings suggest that using self-assessment tools reduces the amount of letters/questions to organizers.

References

1. Admiraal, W., Huisman, B., Pilli, O.: Assessment in massive open online courses. Electron. J. e-Learn. **13**(4), 207–216 (2015)
2. Anderson, A., Huttenlocher, D., Kleinberg, J., Leskovec. J.: Engaging with massive online courses. In: Proceedings of the 23rd International Conference on World Wide Web (WWW 2014), pp. 687–698. ACM, New York (2014). doi:10.1145/2566486.2568042
3. Bali, M.: MOOC pedagogy: gleaning good practice from existing MOOCs. J. Online Learn. Teach. **10**(1), 44–56 (2014)
4. Bayne, S., Ross, J.: The pedagogy of the Massive Open Online Course: the UK view. Report, UK (2013)
5. Dougiamas, M., Taylor, P.C.: Moodle: using learning communities to create an open source course management system. In: Proceedings of the EDMEDIA 2003 Conference, Honolulu, Hawaii (2003)
6. Higgins, C.A., Gray, G., Symeonidis, P., Tsintsifas, A.: Automated assessment and experiences of teaching programming. J. Educ. Resour. Comput. **5**(3), 5 (2005). doi:10.1145/1163405.1163410
7. Hunt, T.: Computer-marked assessment in moodle: past, present and future. In: Proceedings of CAA 2012 Conference (2012)
8. Jerinic, L.: Pedagogical patterns for learning programming by mistakes. In: Proceedings of Conference: Computer Algebra and Dynamic Geometry Systems in Mathematical Education (2012). doi:10.13140/2.1.2720.1609
9. Jordan, K.: MOOC completion rates: the data (2015). http://www.katyjordan.com/MOOCproject.html
10. Kulkarni, C., Wei, K.P., Le, H., Chia, D., Papadopoulos, K., Cheng, J., Koller, D., Klemmer, S.R.: Peer and self assessment in massive online classes. ACM Trans. Comput. Hum. Interact. **20**(6), 33 (2013). doi:10.1145/2505057
11. Papathoma, T., Blake, C., Clow, D., Scanlon, E.: Investigating learners' views of assessment types in Massive Open Online Courses (MOOCs). In: Conole, G., Klobučar, T., Rensing, C., Konert, J., Lavoué, É. (eds.) EC-TEL 2015. LNCS, vol. 9307, pp. 617–621. Springer, Cham (2015). doi:10.1007/978-3-319-24258-3_72

12. Pieterse, V.: Automated assessment of programming assignments. In: van Eekelen, M., Barendsen, E., Sloep, P., van der Veer, G. (eds.) Proceedings of the 3rd Computer Science Education Research Conference on Computer Science Education Research (CSERC 2013), pp. 45–56. Open Universiteit, Heerlen, The Netherlands (2013)
13. Robins, A., Haden, P., Garner, S.: Problem distributions in a CS1 course. In: Tolhurst, D., Mann, S. (eds.) Proceedings of the 8th Australasian Conference on Computing Education, vol. 52, pp. 165–173. Australian Computer Society, Inc., Darlinghurst, Australia (2006)
14. Rodríguez-del-Pino, J.C., Rubio-Royo, E., Hernández-Figueroa, Z. J.: A virtual programming lab for Moodle with automatic assessment and anti-plagiarism features. In: Proceedings of the 2012 International Conference on e-Learning, e-Business, Enterprise Information Systems, & e-Government (2012)
15. Siemens, G.: Massive open online courses: innovation in education? Open Educ. Resour. Innov., Res. Pract. **12** (2013)
16. Singh, R., Gulwani, S., Solar-Lezama, A.: Automated feedback generation for introductory programming assignments. In: Proceedings of the 34th ACM SIGPLAN Conference on Programming Language Design and Implementation (PLDI 2013), pp. 15–26. ACM, New York, (2013). doi:10.1145/2491956.2462195
17. Vihavainen, A., Luukkainen, M., Kurhila, J.: Multi-faceted support for MOOC in programming. In: Proceedings of the 13th Annual Conference on Information Technology Education, pp. 171–176. ACM. (2012). doi:10.1145/2380552.2380603
18. Warren, J., Rixner, S., Greiner, J., Wong., S.: Facilitating human interaction in an online programming course. In: Proceedings of the 45th ACM Technical Symposium on Computer Science Education (SIGCSE 2014), pp. 665–670. ACM, New York (2014) doi: 10.1145/2538862.2538893
19. Widom, J.: From 100 Students to 100,000. ACM SIGMOD Blog (2012)
20. Wilkowski, J., Russell, D.M., Deutsch, A.: 2014. Self-evaluation in advanced power searching and mapping with google MOOCs. In: Proceedings of the first ACM Conference on Learning @ Scale Conference (L@S 2014), pp. 109–116. ACM, New York (2014). doi: 10.1145/2556325.2566241

Assuring Authorship and Authentication Across the e-Assessment Process

Ingrid Noguera[✉], Ana-Elena Guerrero-Roldán, and M. Elena Rodríguez

Universitat Oberta de Catalunya, Barcelona, Spain
{inoguerafr,aguerreror,mrodriguezgo}@uoc.edu

Abstract. As the virtualisation of Higher Education increases, a challenge is growing regarding the use of reliable technologies in blended and on-line learning. Continuous evaluation is being combined with final face-to-face examinations aiming to ensure students' identity. This collides with some universities' principles, such as: flexibility, mobility or accessibility. Thus, it is necessary to develop an e-assessment system to fully virtually assess students and to help teachers to prevent and detect from illegitimate behaviours (i.e. cheating and plagiarism).

The TeSLA project (An Adaptive Trust-based e-assessment System for Learning) is developing the above-mentioned system. During the last eight months the project has been defined in terms of: organisation, pedagogy, privacy and ethics, technologies, quality, and pilot design and evaluation. Currently, the first of the three oncoming pilots is starting.

Keywords: Authorship · Authentication · e-Assessment · Trust · e-Learning process

1 Introduction

Internet has long since opened a new scenario for teaching and learning, allowing people to learn in a more flexible and mobile way. Higher Education institutions are trying to adapt their teaching delivery formats to new demands. ICT (Information and Communication Technologies) are increasingly being incorporated into Higher Education to support the teaching and learning processes by means of diverse pedagogical approaches.

Online education requires a redefinition of the entire teaching-learning processes, from learning activities' design to assessment. Innovations and improvements are implemented in courses design and monitoring, however, more efforts are needed on the e-assessment process. Online, and ICT-oriented, Higher Education institutions currently are quite traditional in terms of assessment. On-site final exams continue being the most common instrument to assess students and to ensure their identity. Research and experiences on e-assessment are proliferating aiming to extend the possibilities of online assessment all along the courses and trying to minimise on-site examinations. In the EHEA[1] (European Higher Education Area), the assessment

---

[1] http://www.ehea.info/pid34247/how-does-the-bologna-process-work.html.

process is the road map for accrediting and ensuring learner competences, thus, when incorporating technologies into the assessment process students' identity and assessment quality have to be guaranteed. Authentication and authorship mechanisms can overcome time and physical barriers and provide assessment with the necessary data to assess students online in a reliable way.

2 The European Higher Education Area Framework

The EHEA has pushed universities to redefine programs in order to facilitate European mobility by means of a common certification system. On the one hand, one of the main changes derived from the EHEA has been the incorporation of competences as the backbone of the programs. For each course there are some expected competences that students will develop. Students have to review provided materials and perform learning activities in order to be assessed and to certify they have developed the competences. This implies a big change, because reading the materials and answering a test do not confirm anymore that a student have learnt what was expected in a course. To be competent means not just to now, theoretically, what was explained in the course but to know how to apply the acquired knowledge. Thus, to teach students to be competent has enormous implications in course design and also on the selection of the assessment model.

On the other hand, assessing by competences forces to move from summative approaches to continuous ones. Assessing students continuously ensures that assessment is not just related to giving marks but it becomes a process from which to learn. Assessing students continuously allow to monitor the students' process and to help them to improve their weaknesses and strengthen their strengths during the course. Currently, this approach is being overcome by formative assessment. Formative assessment promotes the design of interdependent learning activities (i.e. each new activity is based or related to the previous one) and permits to students to improve learning activities during the course.

Competence-based assessment requires diversity of learning activities and assessment instruments. Although test-based exams can be, in some cases, also useful, a combination of more complex learning activities (e.g. projects, study cases, essays, problems resolution, discussions, etc.) and assessment instruments (e.g. portfolios, rubrics, etc.) are required.

2.1 Teaching and Learning Processes

If learning activities and assessment changes when using e-learning platforms, also the roles of the learning agents must change. Competences have to be developed by students, so students become the centre of the learning process, as they are responsible over their learning. Teacher, thus, support them during the course, being always present but remaining in the background as a facilitator. In current Higher Education students are expected to actively participate in their learning and advancing in their competences. Their attitude must be proactive by searching new resources, asking for help and reflecting about their progress. Teachers are expected to carefully design their courses based on competences and to provide individualised and personalised feedback to students continuously.

Learning Management Systems (LMS) and platforms must be flexible enough to allow teachers to monitor diverse learning activities, to assess students in different ways and to facilitate students to actively follow the course and teachers to support students in group and individually.

2.2 Technologies for Enhancing Learning

LMS usually replicated traditional pedagogical approaches (i.e. a teacher giving a lesson) where the teacher transmits information and students receive it and, ideally, learn it just listening and, sometimes, practicing by completing exercises. This is traduced in LMS as a repository of contents where students can download materials and exercises and upload them to be corrected by teachers.

As the pedagogical approaches have been changing, the LMS have evolved from the content-based approach to the activity-based approach. Currently, LMS incorporates communication tools (i.e. forum, microblog, videoconference tools), creation and collaborative tools (i.e. wikis, blogs), assessment tools (questionnaires, peer review, portfolios, rubrics) and even are more easily linked to external tools or social networks (i.e. concept map tools, Twitter, YouTube). These platforms usually have also an assessment system that facilitates teachers to publish the marks during the course and students to consult them. Although these tools facilitate to teach and learn online, there are still challenges on assessing students fully virtually. By using these above-mentioned tools is quite difficult to ensure the identity of students and to prevent from illegitimate behaviours (i.e. cheating and plagiarising). Online learning could be more reliable and credible if authentication and authorship mechanisms were integrated into the learning systems.

2.3 Assessment Models and Learning Activities

E-assessment can be considered as the process where ICT are used for the management of the end-to-end assessment process [1]. In other words, e-assessment deals with methods, processes, and web-based software tools (or systems) that allow systematic inferences and judgments to be made about the learner's skills, knowledge, and capabilities [2].

The e-assessment models are a technological adaptation of the current assessment models. As previously indicated, three assessment models coexist nowadays at European universities: summative assessment, continuous assessment, and formative assessment. Although it is supposed that continuous or formative assessment is the predominant assessment model, in many cases summative assessment or a combination of summative and continuous/formative assessment continues being the dominant option. Furthermore, even existing a large variety of assessable learning activities, activities easy to correct (i.e. tests, multi choice exams) or automatic correction activities proliferate.

All European universities, including 100% online universities, maintain on-site final exams (although there are experiences of online examination). Although students are more and more assessed continuously by performing a variety of learning activities, face-to-face examination is considered the most reliable way to ensure students identity. However, this is not aligned with the common principles of online universities, such as: flexibility, mobility or accessibility. Thus, it is necessary to develop an e-assessment system for Higher Education institutions to fully virtually assess their students. This means for teachers to propose, monitor and assess any type of e-assessment activity being absolutely sure that the students performing such activities are who claim to be. Furthermore, this system has to help teachers to prevent and detect illegitimate behaviours (i.e. cheating and plagiarism).

3 Authorship and Authentication in e-Learning

As previously commented, there is a lack of technologies that incorporates security measures in the e-learning sector. Security measures can promote trust among teachers, students and institution preventing from illegitimate behaviours and incrementing the quality of assessment. Security measures can be defined as learner's authentication on a system when performing or submitting a learning activity, the verification of the authorship of the activity, and the system's protection over the integrity of a submitted activity [5, 7].

The verification of student identity for assessment is based on two key elements: authentication and authorship. Authentication is the process whether someone is declared to be (i.e. login authentication using a password or a digital certificate). Authorship is commonly referred to the identity of the person and is addressed as plagiarism detection in some works. A part from giving a password or the digital certificate, there are other authentication systems that are based on biometric recognition such as: keystroke dynamics, face and voice recognition, fingerprint detection, etc. These systems increase the level of security since this information cannot be just handed away to a third party. Regarding authorship, currently, there are many tools that help to automatically detect plagiarism [3]. However, forensic linguistics techniques are preferred [4], as the authorship of a text is assessed exclusively through comparison of different creations from the same author, like essays [6].

When looking for a scalable approach that can be massively deployed, biometric systems or direct control of personal workspace (keyboard and scriptorium blocks, monitor programs, etc.) become a possibility. However, these systems are usually highly invasive in order to be reliable, which leads to strong privacy problems. Students may feel their personal space has been invaded, watched and questioned by default, undermining the trust between them and the institution.

4 Objectives

Aiming to create an e-assessment system that ensures learners' authentication and authorship, the TeSLA project[2] (*An Adaptive Trust-based e-assessment System for Learning*) is being developed. The consortium is composed by 18 partners and formed by 8 universities (including online and blended universities), 3 quality agencies, 4

---

[2] This project is funded by the European Commission, Horizon 2020 call. It started on 2016 and will finish at the end of 2018. The information regarding the TeSLA project and the consortium can be found at: http://tesla-project.eu/.

research centres, and 3 companies. The main objective of the TeSLA project is to define and develop an e-assessment system, which ensures learners authentication and authorship in online and blended learning environments while avoiding the time and physical space limitations imposed by face-to-face examination. The TeSLA project will support any e-assessment model (summative, continuous and formative) covering teaching and learning processes as well as ethical, legal and technological aspects. TeSLA will offer an unambiguous proof of learners' academic progression, authorship and authentication during the whole learning process.

Some other specific objectives of TeSLA project are:

- Analyse and design the most appropriate learning activities for e-assessment taking into account both, academic requirements to ensure the learning process and the adaptation to a fully online and cross-curricular assessment.
- Improve the e-assessment process by introducing tools and resources in the learning activities that capture learners' data to ensure their authentication and authorship.
- Conduct several pilots of the TeSLA e-assessment system that guarantee the equality of opportunity and respect for diversity in real teaching and learning scenarios while ensuring the authentication and authorship of the learners during the e-assessment processes.
- Hold a set of training workshops, provide guidelines and elaborate learning resources for teachers to show how the TeSLA e-assessment system can be used for enhancing e-assessment processes.

These goals will be achieved by following all the ethical and legal issues related to personal data. All the institutions involved of TeSLA will follow the key principles suggested by the European commission and the regional laws for collecting personal data of teachers, students and institutions. Legal documents like giving consent to participate on pilots are developed to ensure data controller and data processor. Thus, following these principles, TeSLA system will be a reliable and scalable system for any institution which develop its e-assessment processes through the net.

5 Design and Execution of Pilots

5.1 Research Method

The TeSLA project is based on a mixed approach; it follows the principles of an action research methodology combined with a design and creation approach. The research is conducted through three pilots described below. Results from each pilot will improve the next one taking into account the educational and technical aspect as a whole. The mixed approach was selected because it supports innovation as an agile and efficiency manner to create an online system using ICT in order to develop an European e-assessment system from the technical and educational point of view.

The first pilot aims to test the protocol of communication among partners for pilots' execution, the implementation protocol at partner level, and to select the most suitable activities for the e-assessment process at subject level. Seven universities are conducting this small-scale pilot with a sample of approximately 600 students during the first semester of the course 2016–2017.

The second pilot aims to test the modular technologies on an isolated manner in activities, to refine learning activities for e-assessment, and to test and refine the authorship and authentication instruments. Seven universities will conduct this medium-scale pilot with a sample of approximately 3500 students during the second semester of the course 2016–2017. The technologies implemented are:

- *Biometrics*: facial recognition (detects and recognises complex facial information, such as facial expressions), voice recognition (audio structures by segmenting speakers and grouping clusters), and keystroke dynamics (measures how a user writes using the keyboard).
- *Security mechanisms*: digital signature (it is a security mechanism based on a mathematical scheme for demonstrating the authenticity of a digital message or document), and time-stamp (it is a security mechanism based on a sequence of characters or encoded information identifying when an event is recorded by a computer).
- *Document analysis*: plagiarism tools (based on detecting similarities among documents through text matching) and forensic analysis (mechanisms and devices for determining the authorship verification and authorship attribution of written documents).

The third pilot aims to test the full integration of TeSLA system and its scalability, to refine the modular technologies and the European e-assessment framework, and to verity the reliability of the authentication and authorship. This large-scale pilot will be conducted by seven universities in two stages during the course 2017–2018: (1) a pilot during the first semester with a sample of approximately 7000 students, and (2) a pilot during the second semester with a sample of approximately 14000 students.

5.2 Results

Since the beginning of the project (January 2016), tasks from different work packages have been developed in parallel:

- *Pedagogical*: the educational framework has been developed by conducting a literature review about e-assessment, negotiating a common understanding of educational premises and assessment procedures, defining e-assessment patterns, and establishing pedagogical technical requirements.
- *Privacy and ethics*: legal and ethical constraints have been analysed and an agreement document on legal and ethical aspects has been elaborated.
- *Quality assurance*: quality indicators for pilots' evaluation have been created using as a basis the European Standard Guideline (ESG)[3].
- *Technical*: instruments have been analysed to check the transferability to a standardised model, the integration of tools into the diverse learning management systems has been analysed, and technical specifications have been defined in terms of tools, plugins, and security techniques as well as data management plans.

---

[3] http://www.enqa.eu/wp-content/uploads/2015/11/ESG_2015.pdf.

- *Pilots' design*: pilots have been defined following common guidelines (i.e. population, schedules, types of tasks, assessment models). Currently, the first pilot is starting.
- *Pilots' evaluation*: the definition of evaluation measures and questionnaires design have been carried out. Pre pilot questionnaires for teachers and students are being handed out.

As the project evolves some of the above tasks, and new ones, will be conducted for the coming pilots. In following months, we expect to obtain data from pre and post questionnaires and focus groups regarding prior and current experiences on e-assessment, cheating and privacy, and expectations regarding the TeSLA system. The first pilot will also provide meaningful data that will be used to improve the oncoming pilots. As a result of the project, it is expected to obtain an adaptive trust e-assessment system for assuring e-assessment processes in online and blended environments.

6 Conclusion and Future Work

TeSLA is an Innovation Action project which aims to provide to the society with a powerful system to enhance the e-assessment process taking care of the educational premises established in the EHEA but also using technologies being responsible with data protection. Teachers, students, technical staff, legal departments and quality agencies are working together to design, create and test the system for being delivered to the educational society. As the consortium is formed by fully online and blended Higher Education institutions, the pilots will help us to adapt and adjust the system to any kind of learning environment and type of educational model. Taking into account that the first pilot using part of the TeSLA technology is going to start this course, in a near future results will be shared with the community by explaining the authentication and authorship level acquired when teaching and learning process are conducted using the e-assessment system.

References

1. Cook, J., Jenkins, V.: Getting Started with e-Assessment. University of Bath, Bath (2010)
2. Crisp, G.T.: The e-Assessment Handbook. Continuum, London (2007)
3. Diederich, J.: Computational methods to detect plagiarism in assessment. In: 7th International Conference on Information Technology Based Higher Education and Training, Ultimo, New South Wales, pp. 147–154 (2006)
4. Koppel, M., Schler, J.: Exploiting stylistic idiosyncrasies for authorship attribution. In: Proceedings of IJCAI 2003 Workshop on Computational Approaches to Style Analysis and Synthesis, Acapulco, Mexico, pp. 69–72 (2003)
5. Neila, R., Rabai, L.B.A.: Deploying suitable countermeasures to solve the security problems within an e-learning environment. In: Proceedings of the 7th International Conference on Security of Information and Networks, Glasgow, UK (2014)
6. Cook, S.B., Sheard, J., Carbone, A., Johnson, C.: Academic integrity: differences between computing assessments, essays. In: Proceedings of the Koli Calling International Conference on Computing Education Research, pp. 23–32 (2013)
7. Weippl, E.: Security in e-Learning. Advances in Information Security, vol. 16. Springer, USA (2005)

Today - Only TEA and no CAAffee. But Tomorrow?

Rein Prank[✉]

University of Tartu, Institute of Computer Science, Tartu, Estonia
Rein.Prank@ut.ee

Abstract. In this paper we consider two environments for algebraic tasks in Propositional Logic: Truth-Table Checker and Formula Manipulation Assistant. From the very beginning our environments were designed not only for exercise labs but also for assessment of homework and tests. Currently our programs produce detailed data for easy and quick human grading but do not accomplish full Computer Aided Assessment where the output would be a numeric score. We set the goal to take into account five aspects of solutions: what part of the task is solved, errors, hint requests, solution economy or conformity with standard algorithm, and quality of answer. We analyse for different task types what is missing from fully automated grading. We conclude that exploitation of our existing automated solver and integration of some our supplementary programs would enable to provide quite satisfying assessment.

Keywords: Technology Enhanced Assessment · Computer Aided Assessment · Propositional logic · Expression manipulation

1 Introduction

In many books, papers and web pages the term Computer Aided Assessment denotes "assessments completed by the student at a computer without the intervention of an academic" [1]. For instance, the book "Computer Aided Assessment of Mathematics" by C. Sangwin [8] begins with the sentence, "The book examines automatic computer aided assessment (CAA) of mathematics". Many such assessments are conducted using programs that assess the answers but do not touch solutions. Therefore their use in Mathematics is somewhat one-sided. On the other hand, many Technology Enhanced Assessment (TEA) programs are used for assessment of solutions but most of such practical applications contain some participation of human instructors.

Technology Enhanced Assessment software for Mathematics is usually designed for a set of clearly defined task types. A crucial question when designing such software is whether we are able to fully automate the assessment (for a particular task type) or does the program only prepare the data but the assessment should be finalized by an instructor.

For automation we should first decide what features of solutions and answers we are able to assess. For each of these features we should specify the parameters of the solution process, the solution to be submitted at the end and an answer, making it possible to measure this feature. We should implement algorithms for collection of data and computing the values of parameters, and compute the score that corresponds to the values of parameters. By necessity the program should have a user interface for entering the teacher-defined coefficients of grading formulas (weights of subgoals, penalties for error types, prices of using hints, etc.) or even teacher-defined score-counting formulas. In this paper we investigate software implemented at the University of Tartu for exercises and tests in Mathematical Logic. Up to now our assessment is not fully automated and still contains an element of human labour. The paper analyses algebraic task types of the package. We try to ascertain what is missing from satisfactory full automation and how hard it would be to bridge the gap.

Students of our Faculty of Mathematics and Computer Science have used computerised environments for exercises in Mathematical Logic for more than 25 years [2, 7]. We have designed software for truth-table exercises, formula manipulation in Propositional and Predicate Logic, formal proofs in Propositional and Predicate Logic and Peano Arithmetics, Interpreter of Turing Machines, Model Checker for finite models. Over the long period of use we have updated the programs to employ contemporary user interface and server technology, to broaden the spectrum of task types. From the very beginning our environments were designed not only for exercise labs but also for assessment of homework and tests. The programs check what part of the task is solved and what not, diagnose and count errors of different types, measure the size of solution, compare the solution steps to a standard algorithm. However, in most cases our software produces detailed data for easy and quick human grading but does not accomplish full Computer Aided Assessment where the output would be a numeric score.

Section 2 lists the features of solutions and answers that could be taken into account in assessment. Sections 3 and 4 analyse assessment of solutions of Truth-Table and Formula Manipulation tasks. Section 5 presents our conclusions from the analysis.

2 Functional Features of Solutions and Answers

When assessing solutions and answers of most of mathematical tasks, human instructors take into account several aspects:

1. What part of the task is solved (if the solution is incomplete).
2. Errors.
3. Hint requests to the program.
4. Solution economy (in some cases – conformity with algorithm).
5. Quality of answers.

Incomplete solutions can be caused by the time for the test running out, by a lack of skills or solution ideas for some part of the solution, but also by faulty understanding of the goal. The natural way for grading of incomplete solutions is assigning relative weights to subgoals. In case of 'solving another task', it is possible that the student is able to get the right answer if the program gives a corresponding error message. For example, if the student submitted disjunctive normal form instead of conjunctive or vice versa. In our environments such returns belong to category 2, they are counted as Answer errors.

Issue 1 characterises the submitted final solution. Conversely, information about errors is in our algebraic environments collected during the solution dialog. For several reasons we do not use here 'test modes' where solution steps are checked only after submission of an answer. Some types of mistakes (for example, syntactic mistakes) make understanding of entered expressions impossible or ambiguous. Truth-value or equivalence errors give the student an opportunity to change (with purpose or accidentally) the meaning and difficulty of the task (simplifying the expression to be processed, making the equivalent formulas non-equivalent etc.). Our programs for step-based task types check each solution step with regard to possible types of mistakes. If a mistake is detected then the program gives an error message, increases the corresponding error counter and the student should correct the mistake. The environments record only the final versions of solutions that do not contain errors.

If the participants in the test have very different skills or solution of some tasks requires nontrivial ideas then the teacher can decide to allow using hints. Again, the natural and quite universal way of taking the use of hints into account in grading would be assigning costs to possible kinds of hints.

Solution economy characterises the submitted solution but the program can grade it and give the student an opportunity to improve and resubmit the solution (recording thereby the necessary information). For checking of conformity with the 'official solution algorithm' the program can check each step in real time and give messages about deviations. Usually we cannot be sure that every deviation from the 'official' algorithm means that the step is not advantageous. Therefore the message could tell the student what would the standard algorithm do, and the student could have the opportunity to change the step or continue without correction. The program should have counters of different algorithmic errors and/or record the deviation situations. It is also possible to assess only the steps of the submitted final solution because algorithmic mistakes do not change the initial task or the checking opportunities.

Some types of tasks allow correct answers that can be considered having different quality. For example, the formula can have required truth-column or consist of required connectives but allow obvious simplification.

Our package contains two programs for most routine exercises: Truth-Table Environment and Formula Manipulation Assistant. The next two subsections describe and analyse assessment of respective task types.

3 Assessment of Truth-Table Tasks

The truth-table environment is also one of our environments that count the grade for a task. The program supports solution of the following exercise types:

1. Filling the truth-table,
2. Checking of tautology,
3. Checking of satisfiability,
4. Checking of equivalence of two formulas,
5. Checking whether the first formula implies the second,
6. Construction of formula having a given resulting truth-column.

The truth-table environment counts errors of four types: order of operations, truth-values, syntax (only in task type 6) and answer dialog. The teacher program enables assigning a penalty for each type of errors (separately for each task). Our current program gives points only for the finished truth-table tasks. It subtracts the sum of penalties from the points assigned for the task (keeping the score nonnegative).

For solving the tasks of types 2–5 the student first fills the truth-table(s) until the received value(s) allow answering the question. After that the student can switch to answer dialog (Fig. 1). Type 1 contains only the table-filling stage. In tasks of type 6 the student should find and enter a propositional formula that has a given final column of truth-values. It is possible to set an upper bound to the length of the formula and thus eliminate the opportunity of entering a long disjunctive or conjunctive normal form. If the truth-column of the entered formula differs from the given column then the program shows the automatically counted truth-table and the student gets a new attempt to enter the formula. The program detects syntax errors (but we usually assign them zero weight) and attempts with a wrong answer, and applies linear formula for calculation of the score.

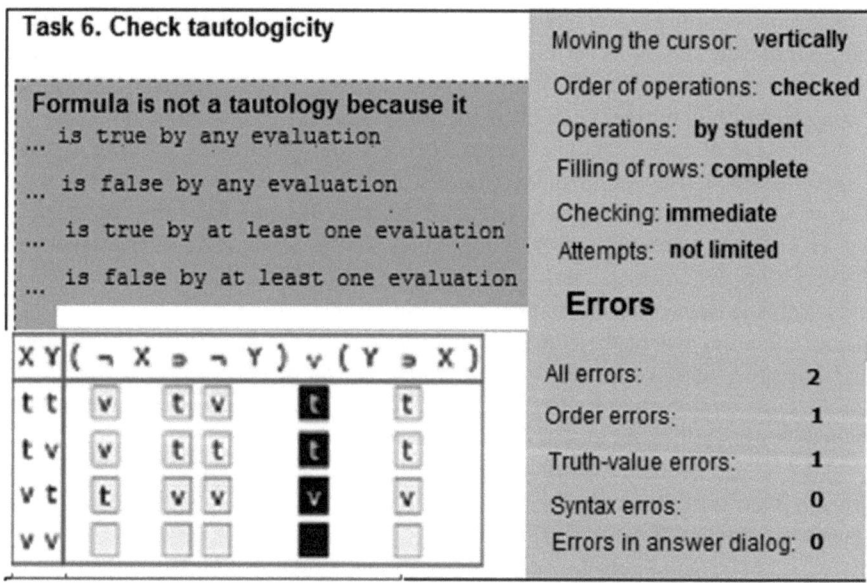

Fig. 1. Solution window for the tautology checking task (original Estonian texts have been replaced with English). The student has filled two rows with resulting values t (true), one with v (false) and is now finishing the answer dialog.

Consider first the most routine task types 1–5 and compare our situation with the desired scheme in Sect. 2.

1. In order to give a nonzero grade for incomplete solutions in task types 2–5, it is possible to consider three stages of solution process and allow the teacher to assign certain partial credit for reaching in the table the stage where the answer can be justified and some credit for giving the right answer (without justification) after that.

2. Using error counters. Working with our current teacher program, the instructor assigns a basic penalty to each error type and the program subtracts the sum of products (error penalty × number of errors) from the task grade. This worked well while our faculty had a small number of students (about 70 per year) and the students were relatively adept. Our typical penalty for order and value errors was 20% of the task grade. The order errors appeared almost exclusively in training labs when the propositional connectives were still new, but not in tests. Most of the students had no truth-value errors in the test and some students had one or two. There could be some incompetent students who had not taken the subject seriously and made mistakes but usually they completely failed the more difficult tasks and had to retake the test after adequate preparation. Now we have three times more participants, many of them are really weak and they continue having errors even after normal amount of training. Our earlier grading principles would give them simply zero points and would not separate students who were able to solve the task but had many errors and students who did not reach the goal at all. In our combined (computer + instructor) assessment we have used some new principles: nonlinear penalty function and 'guaranteed 30% principle'. The latter means that the student gets 30% of points for reaching the goal even when the number of errors is very large. Sometimes we have also left the first error in each task without penalty. Nonlinear counting can be implemented in the teacher program by using a table where the instructor defines penalties for 1, 2, 3, ... errors (the program can propose default values). The other two principles can simply be switched on or off.
3. Our truth-table environment does not give hints. In task types 1-5 the necessary hints are not situation-dependent. When training, the student can consult lecture notes for truth-tables of logical operations, for order of operations and for definitions of tautology etc.
4. The economy issue appears in one aspect of filling a truth-table. In task types 2–5 one of the two possible answers (not a tautology, is satisfiable, not equivalent, ...) can be justified after calculation of only one fitting row, while another result requires filling of the whole table. The students who are not able to guess the answer and the fitting row usually just fill the rows until the result is received (sometimes they even continue after that). We can add to the teacher program the opportunity to assign penalties for calculation of unnecessary values. However, the implementation should be flexible and the teacher should have an opportunity to use particular mechanisms or not. In some tasks the formula can be constructed with the expectation that competent students should be able to guess the values. On the other hand, in case of random generation of formulas, the guessing can have a very variable degree of complexity. It also seems suspicious to take into account the number of needlessly calculated rows. Using the same 'brute force' strategy can give seven superfluous rows in one case and one or even zero in another case. There is one clear symptom of pointless filling the cells - when the student continues calculations after the result is already visible.
5. In tasks of types 1–5 the quality issue is not applicable.

Tasks of type 6 are considerably more difficult and do not have a routine way of solution (if the use of normal forms is excluded). In our tests we ask the students to find a formula that has a 'reasonable' length. For this we use randomly generated columns having 3–6 truth-values true from the eight in the column. With high probability this gives us the column where the normal form is considerably longer than optimal. We have also thought that it would be better if the program enabled random choice from 10–20 fixed columns entered by the instructor. Our tests show that some students fail to find the formula and many others enter the formula that has several connectives more than optimal.

Grading of formula finding tasks has had a much larger human component than many others. The program makes the technical work (checks truth-values of submitted formulas, requires entering of new formula, counts the attempts) but the instructor evaluates the quality of the final formula and finalizes the grading. Consider again the grading issues from Sect. 1.

1. A possible reasonable justification for giving credit for partially solved task could be finding the normal form that corresponds to the given truth-column.
2. The solutions of formula finding tasks do not contain steps. The current program rejects the formulas that have a wrong truth-column and counts unsuccessful submissions.
3. Our current environment does not give hints for type 6. At our exercise labs we show two useful strategies for finding the 'reasonable' formula. One strategy is splitting the column in two halves that correspond to the truth-value of one of the variables and trying then to find patterns of binary connectives, especially of implication and biconditional. For example, the column *tfftfttf* gives us $X\&(Y \sim Z) \vee \neg X\&\neg(Y \sim Z)$ and this can be simplified to $X \sim (Y \sim Z)$. The second strategy is to start with the normal form and simplify it, using distributive and absorption laws and elimination of disjunctions/conjunctions of complementary literals. Hints about useful splitting are to be computed from the initial truth-column. Hints about simplification assume that the student has already entered some formula that has a correct truth-column but the student hopes to simplify it. This means that we have to move type 6 from the truth-table environment to the formula manipulation environment (or integrate them). In fact, the methods that we use in the Formula Manipulation Assistant allow updating the program by diagnosing incomplete simplifications. This would also improve the students' learning already during the exercise labs.
4. Not applicable.
5. The main reason for using human labour in grading of tasks of type 6 has been that the program does not diagnose the qualitative deficiencies of submitted formulas. We want to see students submitting formulas that do not allow obvious simplification using the described above means. Currently this is checked by the human instructor. The mechanism of diagnostics coincides with generating the hints. However, it is possible to use another, more quantitative approach. We can evaluate here just the length of the submitted formula. Currently we have a separate program that finds the formula with the minimum number of connectives for each truth-column. We use the output file of this program (256 lines for 3 variables) for human grading of randomly generated tasks. Adding a corresponding facility to our solution environment can be used for grading of the final formula (replacing the means from the

previous paragraph), for giving the student feedback about the lengthiness of the formula but also to inform the student from the beginning about the ideal length of the formula.

4 Assessment of Expression Manipulation Tasks

The DOS version of our Formula Manipulation Assistant was written by H.Viira-Tamm in 1989–91 [7]. This version implemented exercises with propositional formulas: expression through $\{\neg, \&\}, \{\neg, \vee\}$ or $\{\neg, \supset\}$ and conversion to disjunctive normal form. In 2003 V.Vaiksaar created a Java-version of the program and added task types for Predicate Logic [6]. In 2013 S. Stroom added an Automated Solver for giving hints for the next step. However, the main characteristics of the solution environment are the same as in the first version of the program: the solution dialog with Input and Rule modes, principles of feedback, error diagnosis and classification.

In the Formula Manipulation Assistant (FMA) the student transforms the formula step by step. Each step has two substeps. The first substep is marking of a subformula that will be changed. The second substep depends on the working mode. In the Input mode the replacing formula is entered by the student. In the Rule mode the student chooses a rule from the Rule menu and the program finds the replacing formula automatically. At the first substep the program checks whether the marked part of the formula is a syntactically correct formula and whether this formula is a proper subformula of the whole formula. Syntax errors are usually caused by carelessness (miscounting the parentheses or just a click on the wrong place) but the second type errors usually indicates a misunderstanding of the order of operations. For example, marking of $X \vee Y$ in $X \vee Y \& Z$. is an order error. At the second substep in the Input mode the program checks syntactic correctness of the entered formula, equivalence with the previous formula (for the marked part and for the whole formula) and the addition of parentheses if they are necessary (for example, when de Morgan law was used in $X \& \neg(Y \& Z)$). In the Rule mode the program checks whether the selected rule can be applied to the marked subformula. Figure 2 shows a screenshot of working in the Rule mode. Conversion errors denote selection of an inapplicable rule. In the Input mode the program counts equivalence errors instead.

We consider here only task types of Propositional Logic. Adding quantifiers does not cause additional assessment problems. Conversely, for Predicate Logic we have only the Rule mode. The equivalence of predicate formulas is algorithmically unsolvable and the program cannot check whether the result of entered conversion is equivalent with the previous line.

The Formula Manipulation Assistant implements ten exercise types for Propositional Logic:

1. Three task types on expression of formula through $\{\neg, \&\}, \{\neg, \vee\}, \{\neg, \supset\}$,
2. Conversion to disjunctive/conjunctive normal form (DNF/CNF),
3. Conversion to full disjunctive/conjunctive normal form (FDNF/FCNF),
4. Moving negations inside/outside,
5. Free conversion (the program checks only equivalence at each step).

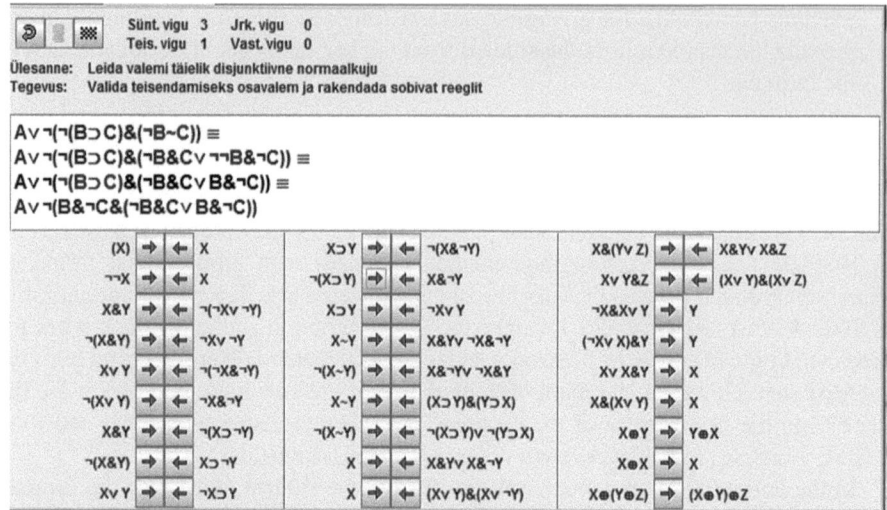

Fig. 2. Solution of FDNF task in Rule mode. The student has eliminated implication and biconditional with 3 syntax errors and 1 rule application error.

In our analyse we refer only to types 1 and 3. In our tests we have used also type 2 but, for the purposes of conversions and error diagnosis, types 2, 4 and 5 are subcases of 1 and 3. The students usually solve in our tests tasks of types 1 in the Input mode (we test the knowledge of conversion rules) and tasks of types 2–3 in the Rule mode (we test the use of normal form algorithm).

Our Formula Manipulation Assistant does not calculate any numeric score for the tasks and a score should be assigned by the instructor. For grading, the program first shows the table with error counters, but the instructor can open the solution log of the task where each step is presented with

(1) initial formula (the marked part has a highlighted background),
(2) selected rule (only in the Rule mode),
(3) resulting formula or error message.

The program contains an automated Solver for types 1–3. It is currently used for hints. The Solver builds its solution as a sequence of steps in the Rule mode.

For comparison of the current situation and possible supplementary features with our vision we use again the list of features from Sect. 2.

1. **Incomplete solutions.** Our program accepts the answers of tasks of types 1–4 only when the requirements are completely satisfied. Otherwise the program counts an answer error, displays an error message and the student can continue the solution. In case of tasks of type 1 some categories of connectives are not eliminated if the student does not know how to express this connective through others. Therefore it could be adequate to divide the credit between three categories of connectives requiring elimination. In normal form tasks there are several possible reasons. Quite frequent is the lack of knowledge about the algorithm. In case of longer formulas

dull execution of the algorithm without using simplification opportunities can result in excessive lengthening of the formula. In the Input mode the work can be stopped by 'missing' conversion rules. It is natural to divide the credit for normal form tasks between stages of the standard algorithm. Such approach can be easily implemented because the completed stage of the algorithm is always recognised by the automated solver (in fact, the necessary syntactical checks are not complex and can also be implemented from scratch).
2. Our program counts errors and they can be used for automated calculation of a score using the same methods as in the case of truth-table tasks.
3. In our test tasks we have suspended hints but in general they can be used in tests too. In our program, generation of hints is based on the automated Solver. Currently the hints are textual, for example, "Express conjunction(s) through negation and implication" or "Move negations inside of conjunctions and disjunctions". However, we could also tell what subformula to mark and what conversion rule to use as the automated solver itself actually works in these terms. Notice that the student should also think in terms of rules regardless of whether the steps are entered in the computer in the Rule or the Input mode. It is possible to divide hints in categories (for example, Marking hints and Rule hints) and give the teacher opportunity to assign costs to each category.
4. Our current formula manipulation environment does not take into account solution economy or conformity with algorithm. However, analyses of test solutions and the length of homework files demonstrate that finding the normal form is a nontrivial task for some students. The students' submit solutions that are significantly longer than those found by the Solver (that uses the standard algorithm). We have experimentally implemented a separate program for checking conformity of solution steps in the Rule mode with the normal form algorithm [3, 4]. It mainly compares the applied rule with the rule that should be used according to the algorithm (the checker also accepts simplification steps). Integration of this checker with the main program allows implementing automated assessment of conformity. There is a small number of solutions where the use of conversions that do not correspond to the algorithm does not increase the length. They can easily be discovered by human checking of cases where the length of solutions and the number of diagnosed differences are in contradiction.

The algorithm for tasks of type 1 is trivial. It is sufficient to 'kill' successively all inadmissible connectives. Expression of conjunction, disjunction or implication through another binary connective and negation requires one conversion step. Biconditional can be eliminated using two or three steps. Typical incompetent step is here expression of conjunction, disjunction or implication using not the goal connective but the remaining third binary operation from $\&$, $\vee$ and $\supset$. This happens when the student does not know the direct conversion formula and seeks for a roundabout way. Another source of supernumerary steps is duplication of connectives requiring elimination when applying an elimination rule to a biconditional. The advantageousness of steps in tasks of type 1 can be easily graded counting the numbers of different connectives in the formula.

In the Rule mode some quite rough evaluation of economy can be received by comparison of the length of the student solution with the solution of the Solver. This approach does not work in the Input mode as the student can enter steps that contain several applications of conversion rules (or even only the answer).

The conformity checking program was written for the solution files of the main program of 2003 that recorded only the successive formulas but not the applied conversion rules. The conformity checking program recognises the applied rule before comparison and we can try applying it to the Input mode solutions too. Scanning the homework files we observed that the students use just the learned conversion rules almost at every step. Practically all naturally arisen exclusions are steps where the student has deleted double negation(s) after some 'main conversion' or has applied 'double' distributive law: $(A \vee B) \& (C \vee D) \equiv A \& C \vee A \& D \vee B \& C \vee B \& D$. Recognition of these two cases can be added to the program. The remaining 'complex' steps tend to be the result of solving the task on paper and entering only some intermediate results. In such situations evaluation of economy, or conformity with the algorithm, is hopeless anyway. Unfortunately we should anticipate that, if we implement evaluation of economy and/or conformity with the algorithm for free-step solutions, then after some time we will get more 'hidden' solutions and we should invent restrictions for this freedom.

5. Our current program checks the answer in task types 1 purely syntactically. Only the formulas that contain inadmissible connectives are refused but a formula with double negations is accepted. It is easy to check applicability of any of the concrete simplification rules but the problem is having task formulations that tell what simplifications should be performed.

Full normal form definitions describe the answer very precisely and there is no space for answers having different quality. In tasks of type 2 we cannot require construction of the minimal normal form. The corresponding algorithms are studied in courses on Boolean functions but not in Logic. Their use is really a special task type. We can require again that some concrete laws (for example, absorption and $X \& Y \vee X \& \neg Y \equiv X$) are applied before submission of the answer.

5 Conclusions

1. As first conclusion of our overview we can tell that already now automated checking of truth-table and equivalence of formulas saves the instructors from the most labour-intensive and error-prone part of the work (not only in our environment but also in many others).
2. Integrating currently existing algorithms and methods in assessment software allows implementing quite diverse testing options: evaluation of incomplete solutions, penalties for errors, using hints during tests, grading of solution economy and conformity with the learned algorithm, quality of answers.
3. Implementation of good assessment requires an automated Solver that builds the solutions in terms of steps that the students are expected to perform. The solver

should be able to work dynamically (starting from the situation created by the student) and should also output an annotated complete solution.
4. Sometimes it could be useful to be able to compute for comparison an ideal solution/ answer that is not required from the student (shortest formula or minimal normal form).
5. In many cases we wrote that the teacher program should enable assigning costs for different accomplishments and penalties for errors. The number of grading variables for a task type can be quite large. However, the program should copy the entered values from one task to another and minimize the work of the instructor.

Some expression manipulation environments use combined working mode where the student should select the (name of) conversion rule but also enter the result of its application. We have created in such style our environment T-algebra [5, 11] for basic school algebra but there are also environment SetSails for expressions of set theory (a consortium of German Universities) [10] and Logex for Propositional Logic (Open University of Netherlands) [9]. A combined mode requires checking whether a step corresponds to selected rule. This is easy if we use a set of formal rewrite rules and require literal application of them, but the situation is much more complex if we want to follow the template of school algebra where the student can simply tell that the used rule is 'Combine like terms'. However, the step can thereby include also multiplication of monomials that in turn includes multiplication of numbers and powers of variables.

References

1. Beeson, M.: Design principles of Mathpert: software to support education in algebra and calculus. In: Kajler, N. (ed.) Computer-Human Interaction in Symbolic Computation, pp. 89–115. Springer, Berlin (1998)
2. Dostalova, L., Lang, J.: ORGANON: learning management system for basic logic courses. In: Blackburn, P., Ditmarsch, H., Manzano, M., Soler-Toscano, F. (eds.) TICTTL 2011. LNCS, vol. 6680, pp. 46–53. Springer, Heidelberg (2011). doi:10.1007/978-3-642-21350-2_6
3. Duží, M., Menšík, M., Číhalová, M., Dostálová, L.: E-learning support for logic education. In: Ariwa, E., El-Qawasmeh, E. (eds.) DEIS 2011. CCIS, vol. 194, pp. 560–568. Springer, Heidelberg (2011). doi:10.1007/978-3-642-22603-8_49
4. Fielding, A., Bingham, E.: Tools for computer-aided assessment. Learn. Teaching Action. **2**(1) (2003). Manchester Metropolitan University, http://www.celt.mmu.ac.uk/ltia/issue4/fieldingbingham.shtml
5. Hettiarachchi, E., Mor, E., Huertas, M.A., Rodríguez, M.E.: A technology enhanced assessment system for skill and knowledge learning. In: CSEDU 2014 (2), pp. 184–191. SciTePress (2014)
6. Holland, G.: Geolog-Win. Dümmler, Bonn (1996)
7. Nicaud, J., Bouhineau, D., Chaachoua, H.: Mixing microworld and cas features in building computer systems that help students learn algebra. Int. J. Comput. Math. Learn. **5**, 169–211 (2004)
8. Prank, R.: Trying to cover exercises with reasonable software. In: Second International Congress on Tools for Teaching Logic, pp. 149–152. University of Salamanca (2006)

9. Prank, R.: Software for evaluating relevance of steps in algebraic transformations. In: Carette, J., Aspinall, D., Lange, C., Sojka, P., Windsteiger, W. (eds.) CICM 2013. LNCS, vol. 7961, pp. 374–378. Springer, Heidelberg (2013). doi:10.1007/978-3-642-39320-4_32
10. Prank, R.: A tool for evaluating solution economy of algebraic transformations. J. Symbolic Comput. **61–62**, 100–115 (2014)
11. Prank, R., Issakova, M., Lepp, D., Tõnisson, E., Vaiksaar, V.: Integrating rule-based and input-based approaches for better error diagnosis in expression manipulation tasks. In: Li, S., Wang, D., Zhang, J.-Z. (eds.) Symbolic Computation and Education, pp. 174–191. World Scientific, Singapore (2007)
12. Prank, R., Vaiksaar, V.: Expression manipulation environment for exercises and assessment. In: 6th International Conference on Technology in Mathematics Teaching, Volos-Greece, October 2003, pp. 342–348. New Technologies Publications, Athens (2003)
13. Prank, R., Viira, H.: Algebraic manipulation assistant for propositional logic. Comput. Logic Teaching Bull. **4**, 13–18 (1991)
14. Sangwin, C.J.: Assessing elementary algebra with STACK. Int. J. Math. Educ. Sci. Technol. **38**, 987–1002 (2007)
15. Sangwin, C.: Computer Aided Assessment of Mathematics. Oxford University Press, Oxford (2013)
16. LogEX. http://ideas.cs.uu.nl/logex/
17. SetSails. http://sail-m.de/sail-m.de/index.html
18. T-algebra. http://math.ut.ee/T-algebra/

A Hybrid Engineering Process for Semi-automatic Item Generation

Eric Ras[✉], Alexandre Baudet, and Muriel Foulonneau

IT for Innovative Services, Luxembourg Institute of Science and Technology,
5, Avenue des Hauts-Fourneaux, 4362 Esch-sur-Alzette, Luxembourg
{eric.ras,alexandre.baudet,muriel.foulonneau}@list.lu

Abstract. Test authors can generate test items (semi-) automatically with different approaches. On the one hand, bottom-up approaches consist of generating items from sources such as texts or domain models. However, relating the generated items to competence models, which define required knowledge and skills on a proficiency scale remains a challenge. On the other hand, top-down approaches use cognitive models and competence constructs to specify the knowledge and skills to be assessed. Unfortunately, on this high abstraction level it is impossible to identify which item elements can actually be generated automatically. In this paper we present a hybrid process which integrates both approaches. It aims at securing a traceability between the specification levels and making it possible to influence item generation during runtime, i.e., after designing all the intermediate models. In the context of the European project EAGLE, we use this process to generate items for information literacy with a focus on text comprehension.

Keywords: Automatic item generation · Formative assessment · Information literacy · Cognitive model · Competence model

1 Introduction

The manual creation of assessment items by humans suffers a number of limitations. It is not a scalable process since the time involved for creating individual items is very significant and hence, authoring a large amount of items for a continuous formative assessment approach is challenging. The creation of items is perceived as an "art" and leads to inconsistencies amongst item developed by different authors for the same tasks. Finally the cost of creating items manually has appeared prohibitive for the creation of large item banks [1]. As a result Automatic Item Generation (AIG) has gained interest for the last years, mainly in the language learning and the mathematical domains.

In the European project EAGLE[1] we aim to support the development of assessment items to support formative assessment in the context of the public administration.

EAGLE's learning approach is based on so-called open educational resources (OER). Such OERs are created by employees of the public administrations for the

---

[1] http://www.eagle-learning.eu/.

purpose of workplace learning and knowledge sharing. They can have different formats such as text, video, audio, presentation or interactive content. The EAGLE platform contains also several community tools such as a Wiki and blogs.

We propose using a process that can support the generation of items from the definition of construct maps in the domain of information literacy as well as from text-based OERs and Wiki pages. We call this process a hybrid AIG approachbecause it uses a top-down engineering approach, i.e., starting from a competence construct as well as bottom-up, i.e., processing a text-based resources available on the EAGLE platform in order to influence the generation of the items at runtime.

Section 2 elaborates on existing research work in the domain of automatic item generation. Section 3 describes the hybrid engineering workflow and the underlying models used in the EAGLE project. Section 4 provides an example of item generation in the content of text comprehension, which is a basic skill in information literacy.

2 Related Work on Automatic Item Generation

Typically, we distinguish between systematic bottom-up and systematic top-down approaches for AIG.

AIG top-down approaches, such as the one described by Luecht [2] follow a systematic engineering process: knowledge and skill claims on different performance levels reflecting different proficiency claims are described in so-called construct maps. In Assessment Engineering (AE), Luecht proposes a scale of ability which describes how well the assessed person masters tasks in a specific domain. The next refinement step specifies a task model, which defines a unique combination of skills (i.e., procedural skills) applied to knowledge objects within specific ranges/levels of the construct map. The last step is to design item templates. An item template is an expression of those requirements (knowledge and skills) with the inclusion of item content and the context or domain. Items are varied by means of variables in the stem, options and auxiliary information [1, 3, 4]. A drawback of this approach is the level of flexibility it allows to generate items when the context is changing: By following the AE approach, the designer reaches a detailed specification level by defining the corresponding variables in an item template. The problem of these approaches is that a change of auxiliary information in the item triggered by a change of the assessment domain (e.g. the knowledge domain where the formative test is used) can only be implemented in the item template when its design is revised manually. This is an effort intensive and cumbersome activity for an AIG designer in particular when many item templates needs to be changed.

We call those AIG approaches *bottom-up* when a so-called knowledge excerpt of a source is used to build an item. A source can be a text or a domain model. Using a text as input, linguistics features and syntactical information can be used to influence item generation. A domain model provides knowledge excerpts such as key concepts, references, relationships between two concepts in a clause, etc.

Bottom-up strategies can represent very different approaches with various levels of automation and various sources to be used for AIG. While Gierl et al. [5] only use manually created templates and variables generated from the definition of mathematical equations (e.g., x is an integer between 1 and 10), Karamanis et al. [6] and Moser et al. [7] generate questions from texts. Foulonneau et al., Linnebank et al., Liu et al. and Papadopoulos et al. use semantic models in domains of medicine [8–11], history and environment, and computer science to generate items. Finally, Zoumpatianoset al. [12] suggests generating choice questions from rules using a *Semantic Web Rule Language* (SWRL).

Karamanis [6] assesses the quality of questions obtained through AIG by asking experts to identify the questions which could be delivered without edit. Nevertheless, the quality of the generated items mainly depends on linguistic features [13] and distractors (i.e., wrong answer options) effectiveness [8] rather than on the quality of the items to adequately measure a construct (i.e., the knowledge or skill that is being assessed). Most of these approaches are triggered by the source rather than by an educational objective or a higher level competence construct, which describes the knowledge and required skills to be tested. Not having these specifications makes it very difficult to argument about validity for test-based inferences and to decide about the difficulty level of the generated items.

Today, the top-down and the bottom-up approaches are complementary but unfortunately not integrated. In order to cope with their limitations we propose a hybrid approach which integrates both. On the one hand, specification models on different levels of abstraction allow a step-wise reduction of complexity, a good traceability between the different models, and a higher level of reuse. On the other hand the process leads to a sufficient detailed specification of items templates for automatic item generation. The use of context specific information allows to influence the generation of items during runtime, i.e., when the generation process is triggered by a user.

3 The Hybrid Engineering Process for AIG

The AIG process in embedded in an eAssessment workflow (see Fig. 2) where different actors are involved and where information (e.g., models, specifications, items, tests) is passed from one phase to the other using different tools. In EAGLE, the AIG process is embedded into the workflow of our TAO® platform [14] (Fig. 1).

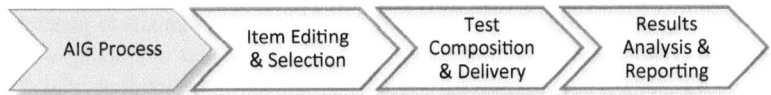

Fig. 1. eAssessment workflow

AIG is the automatic or semi-automatic creation of test items. Figure 2 shows the different modelling steps to derive the models for generating items. The process is adapted from the engineering process of Luecht [2]. We consider the AIG process as a process with five main steps. Up to now, the process has ICT-support for the last two phases, whereas the first three phases are still human-based design steps. The step *Item Generation* is fully automated. In the following, we will explain the different modeling steps. The square brackets emphasize that the step is not mandatory.

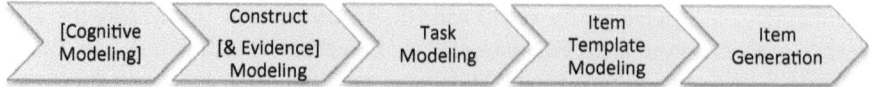

Fig. 2. AIG process derived from [15]

On the higher level of abstraction a cognitive model is defined. Unfortunately, there is no standard of how cognitive models are developed and used to guide item development [16]. Leighton and Gierl [16] state that "cognitive models can be used to support the identification and relationship among the knowledge, skills, and competences that are the focus of testing [...], and ultimately inferences about student's mastery of the knowledge and skills". Typically test developers rely on expert opinions and test specifications. Nevertheless, *cognitive models of task performance* derived from empirical observations should be used for AIG. They are the only type of cognitive models that are suitable to make empirically-based inferences about the test taker's knowledge and skills.

For our purpose we require that a cognitive model of task performance has to provide information about the sequence of cognitive processes as well as constraints and their inter-dependencies. The cognitive processes have to be formulated in a context-free manner and independent of the proficiency level.

We use a diagrammatic representation because it reveals to be most useful to describe inter-dependencies between cognitive processes. The reason in that our AIG process is using additional abstraction layers to refine the cognitive dimension. The next abstraction layer are the construct maps. They support the items designer to derive knowledge and skills claims from the cognitive model.

Wilson [17] describes a construct map as "a well thought out and researched ordering of qualitatively different levels of performance focusing on one characteristic". In EAGLE, for each information literacy competency, a construct map has been developed, which is broken down into knowledge and skills claims. A competence represents the ability to combine knowledge and skills to perform actions. Therefore, a construct map has different *levels of proficiency*. Optionally to the skills and knowledge specification, concrete performance-based evidences can be provided in addition to justify when a level has been reached (see also [2]).

Unfortunately, a construct map still doesn't provide enough details to generate items, therefore a *task model* defines combinations of cognitive activities interacting with knowledge objects [2]. The model defines the knowledge and skill requirements that should be implemented in the design of an item template (see also Lai and Gierl [18]). At this level, the hybrid approach differs from the other AE approaches summarized earlier. Typically, a task models specifies also the content and context dimension since it has an impact on the cognitive complexity of a task [2]. In our hybrid approach this information is specified on the item template level. It leads to a higher level of reuse on the task model level, i.e., a task model contains *task expressions*. Such a set of expressions is linked to one claim of the construct map. One single expression can be either atomic or derived. An *atomic expression* specifies a cognitive process and its interaction with a knowledge object on the lowest level of granularity. As a basic taxonomy of cognitive processes we refer to the cognitive processes of Anderson and Krathwohl [19]. A knowledge object can be for example a fact, a concept or relation from a domain model, or a knowledge excerpt from source (e.g., text). Atomic processes can be combined to form *composed expressions*. A composed expression allows formulating more complex interactions involving combinations of more than one cognitive process and several knowledge objects. A task model is used as a basis for the creation of multiple item templates.

Items are generated from an item template. An item template is a representation of an item with the specification of elements (variables) that can be modified under certain conditions in order to generate multiple isomorphic items from the same template. The generated items will have the same psychometric properties (e.g. difficulty). The variables allow to vary the stem, the options, the auxiliary elements of an item and also the external resources (e.g., domain model, Web resource, data base, text, etc.), which is used to resolve the variables. When the generation process is triggered by the user each variable is resolved by the generation process. An item template refers to a specific interaction type (e.g., match, cloze, MCQ, etc.). The advantage of using an item template is that the psychometric properties do not have to be re-calculated for each item generated from the same template. Only a "parent item", used as a representative of all items generated with the same template is in this case calibrated, with the possibility to determine the psychometric properties of all items generated from the same parent item (see also [20]).

The next section will follow this process and will provide an example for each step.

4 Example – From a Construct Map to Generated Items

The first model to be designed is the cognitive model. In the context of the EAGLE project, we used the ACRL framework [21] as a basic to structure the different relevant constructs for information literacy. The ACRL framework is comprehensive and describes information literacy related competences in so-called standards. To limit the scope for this paper we decided to focus on the standard *Evaluate information and its sources critically*. This is a typical situation where employees of the public administration process textual information such as internal process descriptions, regulations, law texts or inquiries received from citizens by email.

The cognitive model shown in Fig. 3 was modelled usingthe Business Process Model and Notation (BPMN) [22], which is known for designing business processes. The elements of a BPMN model are well defined and allow specifying flows of cognitive actions. A cognitive model is typically composed of states, processes, and possible paths through the model. Arrows describe the sequence of cognitive processes as they are typically performed when solving a task, in this case understanding and evaluating a text.

As mentioned in the previous section, a cognitive model does not represent different proficiency levels. Nevertheless, some processes are mandatory to proceed further (e.g., 1, 2, 3 in Fig. 3) or alternative options are possible (4, 5, 6). The operator "X" is used as an exclusive choice of options whereas the operator "+" allows cognitive process to take place in parallel ((Group: 1-2-3-4-5-6), 8, 9). For each proficiency level a construct map is developed. In this example we elaborate the construct map for the high proficiency level. Each

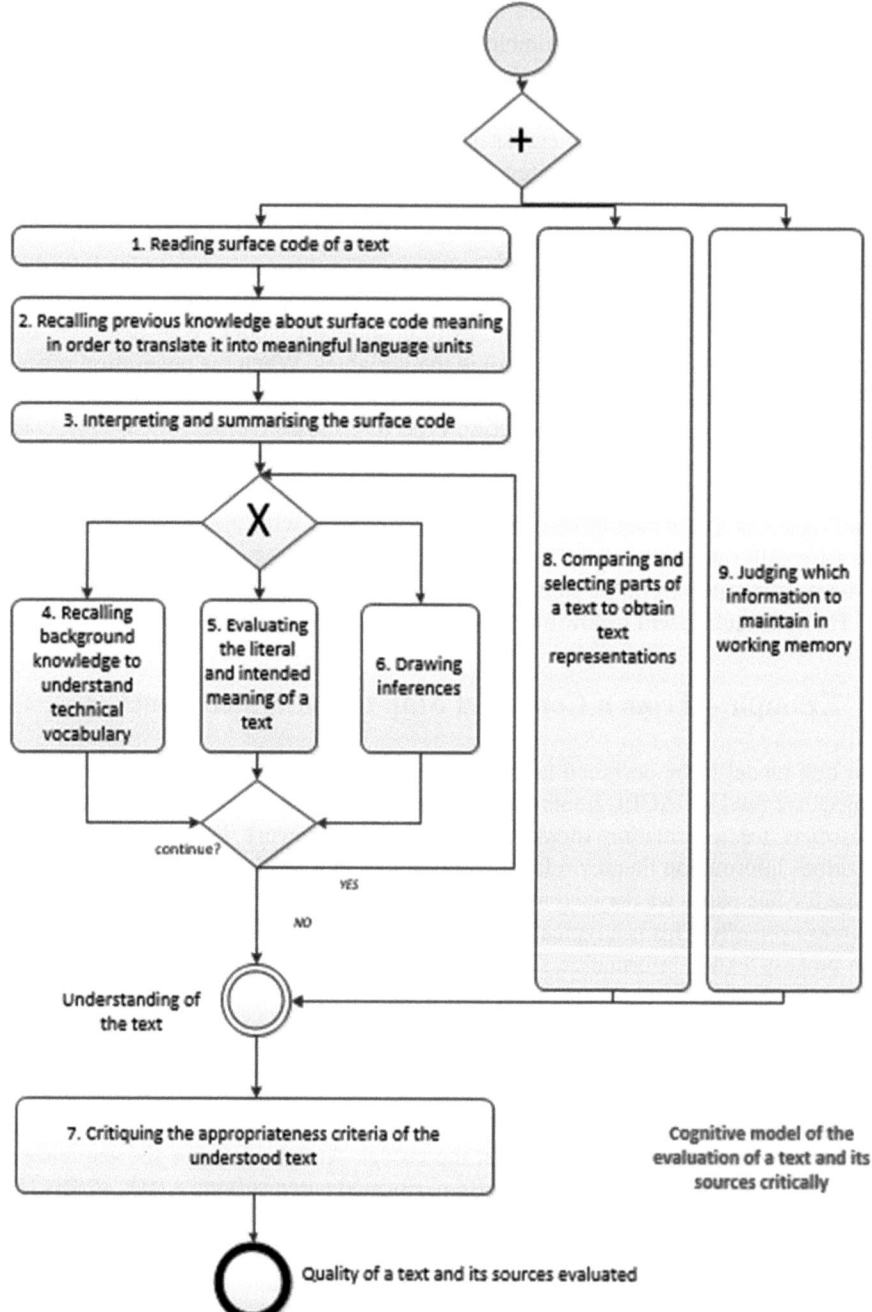

Fig. 3. Cognitive model of the *evaluation of a text and its sources critically*

knowledge or skill component is related to at least one cognitive process and is numbered for traceability (Table 1).

Table 1. Example of a construct map on the high proficiency level

Type	Description	Reference cog. model
K1	Knowledge of the surface code meaning (wording and syntax principles)	2
K2	Knowledge of specific vocabulary	4
S1	Able to draw bridging inferences: conclusions drawn to increase coherence between the current and preceding parts of a text	6
S1	Able to draw elaborative inferences: inferences based on our knowledge of the world that involve details to the text that is being read	6
K3	Knowledge of the literal and intended meaning characteristics of a text (irony, humour, etc.)	5
K4	Knowledge of the assessment's criteria of an information and its sources (author's credibility, accuracy, publication date, etc.)	7

Next, we present a task model for the component K2, which is linked to the cognitive process "Recalling background knowledge to understand technical vocabulary". Each component has one task model. A task model has one to several task model expressions. In our case three expressions are formulated:

All expressions are atomic. Vocabulary learning can take place on different levels of cognition. On the lowest level of remembering a learner can recall a synonym of a presented term and on the understanding level a learner can match different terms with a definition by inferring similarities between the terms and the text of a definition (Fig. 4).

```
Task model title: Knowledge of specific vocabulary
     Task model expression 1: recall(term, term synonym)
     Task model expression 2: compare(term definition, term)
     Task model expression 3: infer/match(text entity, term)
```

Fig. 4. Example of task model for *Knowledge of specific vocabulary*

A single task model expression (TME) can be implemented using different item templates. TME1 is implemented by an MCQ item with several options. TME2 consists in identifying a term from its definition. In this task, the public administration employee is invited to choose between multiple options in an MCQ and identify the correct concept to which the definition applies. For TME3 a cloze item template is used.

The XML-based item template has different sections. All sections are shown by excerpts from a MCQ item template (see Fig. 5). The metadata section describes the references to the other models and also the item type. The second section contains the body of the item with the elements: stem, key, options, and auxiliary information. The advantage or our item templates is that the specification of the item body is fully QTIv2.1

compliant and the generated items can therefore be imported into any eAssessment platform that can import QTI items. The next section is dedicated to the variables which are used to define the variants of the items. Each variable refers to a process that resolves the variable during runtime, i.e., the variable is replaced with content for this element of the item.

```xml
<?xml version="1.0" encoding="UTF-8"?>
<template>
    <metadata>
        <identifier>http://list.lu/assessment/itemTemplate/ChoiceTermDefinition1</identifier>
        <cognitiveModel>http://list.lu/assessment/cognitiveModel/1</cognitiveModel>
        <cognitiveProcess>http://list.lu/assessment/cognitiveProcess/1</cognitiveProcess>
        <constructMap>http://list.lu/assessment/constructMap/1</constructMap>
        <constructMapComponent>http://list.lu/assessment/constructMapComponent/1</constructMapComponent>
        <taskModel>http://list.lu/assessment/taskModel/1</taskModel>
        <taskModelElement>http://list.lu/assessment/taskModelElement/findTermFromTermDefinition</taskModelElement>
        <interactionType>choiceInteraction</interactionType>
        <keyVariable>http://list.lu/term/keyVariable</keyVariable>
        <correctResponseVariable>http://list.lu/term/correctResponseVariable</correctResponseVariable>
        <distractorVariable>http://list.lu/distractor</distractorVariable>
        <correctResponseAttributionMode>RANDOM</correctResponseAttributionMode>
        <distractorAttributionMode>RANDOM</distractorAttributionMode>
    </metadata>
    <layer>
        <assessmentItem xmlns="http://www.imsglobal.org/xsd/imsqti_v2p0" xmlns:xsi="http://www.w3.org/2001/XMLSchema-instance"
            xsi:schemaLocation="http://www.imsglobal.org/xsd/imsqti_v2p0 imsqti_v2p0.xsd" ...>
            ...
            <itemBody>
                <choiceInteraction responseIdentifier="RESPONSE" shuffle="false" maxChoices="1">
                    <prompt>Which term corresponds to the following definition {http://list.lu/termDefinition}?</prompt>
                    <simpleChoice identifier="{http://list.lu/optionCode1}">{http://list.lu/optionLabel1}</simpleChoice>
                    ...
                </choiceInteraction>
            </itemBody>
            ...
        </assessmentItem>
    </layer>
    <variableDefinitions>
        <variable>
            <identifier>http://list.lu/termDefinition</identifier>
            <language>en</language>
            <initialisation>http://list.lu/process1</initialisation>
            ...
        </variable>
        ...
    </variableDefinitions>
    <processes>
        <process>
            <identifier>http://list.lu/process1</identifier>
            <type>SemanticRetrievalProcess</type>
            <datasource>http://list.lu/onDemandTextOntology</datasource>
            <query>
                <query><![CDATA[
                PREFIX : <http://www.list.lu/onDemandTextOntology#>

                SELECT DISTINCT ?http://list.lu/termURI (str(?termLabel) as ?http://list.lu/termLabel )
                ?http://list.lu/termDefinitionURI
                    (str(?termDefinitionLabel) as ?http://list.lu/termDefinitionLabel )
                WHERE
                {
                    ?http://list.lu/termURI a :Word;
                                            :content ?termLabel;
                                            :hasDefinition ?http://list.lu/termDefinitionURI.
                    ?http://list.lu/termDefinitionURI a :Sentence;
                                                      :content ?termDefinitionLabel.
                }
                LIMIT 1]]>
                </query>
            </query>
            <input>
                <inputIdentifier>http://list.lu/conceptURI</inputIdentifier>
            </input>
            <outcome>
                <outcomeIdentifier>http://list.lu/correctResponse</outcomeIdentifier>
            </outcome>
        </process>
        ...
    </processes>
</template>
```

Fig. 5. Item type template MCQ for TMS2 *compare (term definition, term)*

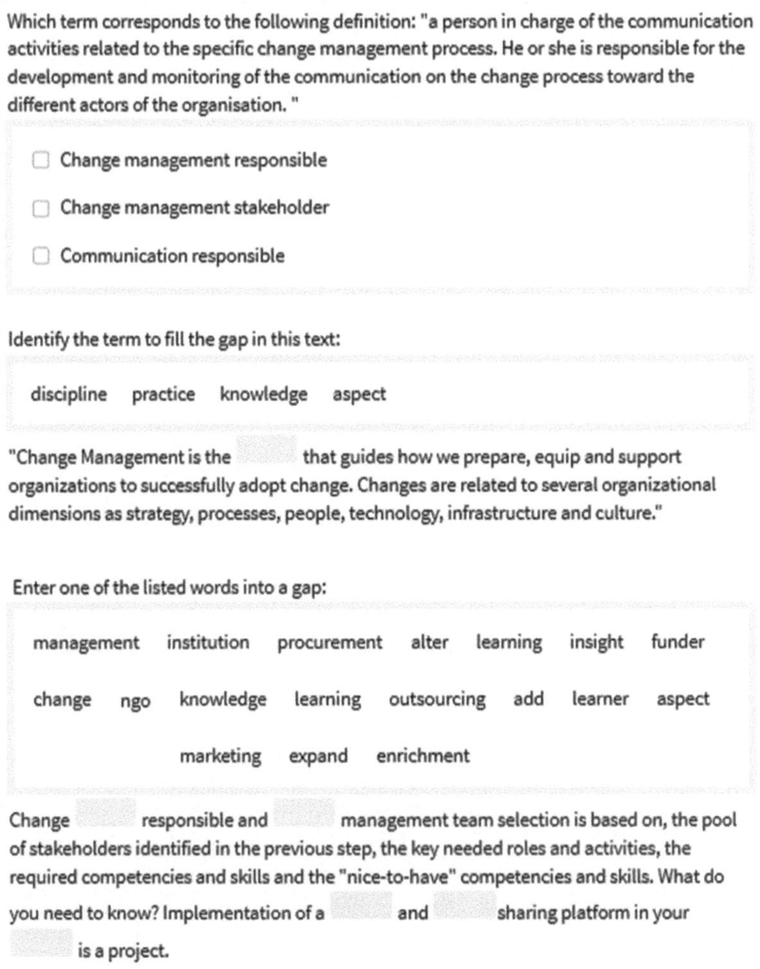

Which term corresponds to the following definition: "a person in charge of the communication activities related to the specific change management process. He or she is responsible for the development and monitoring of the communication on the change process toward the different actors of the organisation. "

☐ Change management responsible

☐ Change management stakeholder

☐ Communication responsible

Identify the term to fill the gap in this text:

discipline   practice   knowledge   aspect

"Change Management is the ____ that guides how we prepare, equip and support organizations to successfully adopt change. Changes are related to several organizational dimensions as strategy, processes, people, technology, infrastructure and culture."

Enter one of the listed words into a gap:

management   institution   procurement   alter   learning   insight   funder

change   ngo   knowledge   learning   outsourcing   add   learner   aspect

marketing   expand   enrichment

Change ____ responsible and ____ management team selection is based on, the pool of stakeholders identified in the previous step, the key needed roles and activities, the required competencies and skills and the "nice-to-have" competencies and skills. What do you need to know? Implementation of a ____ and ____ sharing platform in your ____ is a project.

Fig. 6. Screenshot of generated items in TAO®

All resolving processes are defined in the last section of the template. In this example the variable *term Definition* used in the prompt (i.e., stem) is retrieved by executing a semantic query to a domain model, which is also determined in the template. In EAGLE, this domain model is extracted automatically from a resource, in this case a text-based OER.

After resolving all the variables, each item is stored in the TAO® platform. In EAGLE the test designer has the possibility to manually select items and compose a test or to let the AIG process compose automatically a test by using specific criteria: The user can for example select from three difficulty levels and the user can also define which types of items should be used in the test.

The following figure shows one page of a test composed of an MCQ, cloze, and gap match item for the construct *knowledge of specific vocabulary*. The items were generated from OERs of the EAGLE platform on the topic of change management in the public administration (Fig. 6).

5 Conclusion and Future Work

In this paper we have described the AIG process and as well as its implementation inspired by the frameworks proposed by Luecht and Gierl et al. [1] in order to support formative assessment for public administrations. The different models have been created for EAGLE on the different levels. We have proposed notations as well as templates to specify the models on the different abstraction levels. These abstraction levels reduce the complexity for item designers and hence bridge the gap between a high level cognitive model and the fine grained item template specifications. The example demonstrates that the traceability between the generated items and the constructs is possible.

Unfortunately, most AIG approaches do not cover all types of items but mainly focus on generating multiple choice questions (MCQ). The use of variables in multiple item templates and the fact that all templates are QTI compliant, don't limit the approach to specific item types. The limits reside on the resolving process of the variables defined in the template. If the resource that is used as input does not contain the information to resolve the variable, this means that the item cannot be generated. Therefore, AIG strongly depends on the domain model or the knowledge excerpts that are extracted from a given resource (e.g., text, Wiki page, OER).

Specifying the models for AIG requires an interaction between the domain experts/ instructors and the item bank creator to ensure that the final item bank will contain appropriate items and the tests will be valid. It is important to decrease the effort necessary for the creation of item templates because a complex task model can lead to many item templates. This can be addressed by increasing the level of automation of the domain model extraction with semantic annotation mechanisms (e.g., [20]) and authoring guidelines (e.g., editing models such as the Text Encoding Initiative [23], Wiki infobox templates [21], and semantic Wikis) and the semi-automatic creation of templates from the ontology based on ontology patterns.

Another current drawback is that we didn't integrate yet the modelling of the cognitive model, construct and task model into an IT-supported workflow. Therefore in the future, we will also choose appropriate tools which can exchange machine readable versions of the models. This allows in addition a systematic verification of the specifications, i.e., validity, completeness, consistency, etc.

A major challenge is to evaluate the quality of the generated items and also the AIG-related modelling process. Therefore, we aim to conduct a comparative case study where test design experts either follow the AIG process as elaborated in this paper or where they follow the traditional process of item design. In addition, we will derive quality indicators from the elements of the items (e.g. semantic similarity between distractor and key) and characteristics of the resources (e.g., reading complexity of a text) in order to develop an automatic item features analysis tool. The tool will support the classification of items in the item bank and hence the composition of suitable tests for the different proficiency levels and learning needs. This will lead to an extension of the model chain with a quality dimension, i.e., while we define the other models, we will infer quality related information to build a quality model. Different quality mechanisms and algorithms will be developed to derive psychometric characteristics for the items. These quality mechanisms will certainly not replace the manual validation process performed by humans. In future experiments, items generated should be compared with "traditional items" to evaluate the quality of our automatic generated items.

Finally, the implemented feedback mechanism is simple, in the sense that TAO® provides only corrective feedback and no comprehensive feedback (e.g., prompt, hint, tutorial feedback, etc.). We have started the design of a feedback system for AIG based on incremental user models which are derived from users' behavior in tests [24].

Acknowledgements. This work was carried out in the context of the EAGLE funded by the European Union's Seventh Framework Programme for research, technological development and demonstration under grant agreement No. 619347.

References

1. Gierl, M.J., Haladyna, T.M.: Automatic Item Generation - Theory and Practice. Routledge, New York (2013)
2. Luecht, R.M.: An introduction to assessment engineering for automatic item generation. In: Gierl, M.J., Haladyna, T.M. (eds.) automatic item generation. Routledge, New York (2013)
3. Gierl, M.J., Lai, H.: Using weak and strong theory to create item models for automatic item generation. In: Gierl, M.J., Haladyna, T.M. (eds.) Automatic Item Generation. Routledge, New York (2013)
4. Gierl, M.J., Lai, H.: Generating items under the assessment engineering framework. In: Gierl, M.J., Haladyna, T.M. (eds.) Automatic Item Generation. Routledge, New York (2013)
5. Gierl, M.J., Lai, H.: Methods for creating and evaluating the item model structure used in automatic item generation, pp. 1–30. Centre for Research in Applied Measurement and Evaluation, University of Alberta, Alberta (2012)
6. Karamanis, N., Ha, L.A., Mitkov, R.: Generating multiple-choice test items from medical text: a pilot study. In: Fourth International Conference Natural Language Generation Sydney, Australia, pp. 111–113 (2006)
7. Moser, J.R., Gütl, C., Liu, W.: Refined distractor generation with LSA and Stylometry for automated multiple choice question generation. In: Thielscher, M., Zhang, D. (eds.) AI 2012. LNCS, vol. 7691, pp. 95–106. Springer, Heidelberg (2012). doi:10.1007/978-3-642-35101-3_9
8. Foulonneau, M., Groués, G.: Common vs. expert knowledge: making the semantic web an educational model. In: 2nd International Workshop on Learning and Education with the Web of Data (LiLe-2012 at WWW-2012 Conference), vol. 840, Lyon, France (2012)
9. Linnebank, F., Liem, J., Bredeweg, B.: Question generation and answering. DynaLearn, Deliverable D3.3, EC FP7 STREP project 231526 (2010)
10. Liu, B.: SARAC: a framework for automatic item generation. In: Ninth IEEE International Conference on Advanced Learning Technologies (ICALT2009), Riga, Latvia, pp. 556–558 (2009)

11. Papadopoulos, Pantelis M., Demetriadis, Stavros N., Stamelos, Ioannis G.: The impact of prompting in technology-enhanced learning as moderated by students' motivation and metacognitive skills. In: Cress, U., Dimitrova, V., Specht, M. (eds.) EC-TEL 2009. LNCS, vol. 5794, pp. 535–548. Springer, Heidelberg (2009). doi:10.1007/978-3-642-04636-0_49
12. Zoumpatianos, K., Papasalouros, A., Kotis, K.: Automated transformation of SWRL rules into multiple-choice questions. In: FLAIRS Conference, Palm Beach, FL, USA, vol. 11, pp. 570–575 (2011)
13. Mitkov, R., Ha, L.A., Karamanis, N.: A computer-aided environment for generating multiple-choice test items. Nat. Lang. Eng. **12**, 177–194 (2006)
14. TAO platform. https://www.taotesting.com/. Accessed 16 Dec 2016
15. Foulonneau, M., Ras, E.: Automatic Item Generation - New prospectives using open educational resources and the semantic web. Int. J. e-Assessment (IJEA) **1**(1) (2014)
16. Leighton, J.P., Gierl, M.J.: The Learning Sciences in Educational Assessment - The role of Cognitive Models. Cambridge University Press, New York (2011)
17. Wilson, M.: Measuring progressions: assessment structures underlying a learning progression. Res. Sci. Teach. **46**, 716–730 (2009)
18. Lai, H., Gierl, M.J.: Generating items under the assessment engineering framework. In: Gierl, M.J., Haladyna, T.M. (eds.) Automatic Item Generation. Routledge, New York (2013)
19. Anderson, L.W., Krathwohl, D.R.: A Taxonomy for Learning, Teaching, and Assessing: a Revision of Bloom's Taxonomy of Educational Objectives. Longman, New York (2001)
20. Sonnleitner, P.: Using the LLTM to evaluate an item-generating system for reading comprehension. Psychol. Sci. **50**, 345–362 (2008)
21. ACRL framework (2016). http://www.ala.org/acrl. Accessed 6 Dec 2016
22. Object Management Group Business Process Modeling and Notation v2 (BPMN). http://www.bpmn.org/. Accessed 19 Dec 2016
23. Text Encoding Initiative (TEI) (2016)
24. Höhn, S., Ras, E.: Designing formative and adaptive feedback using incremental user models. In: Chiu, D.K.W., Marenzi, I., Nanni, U., Spaniol, M., Temperini, M. (eds.) ICWL 2016. LNCS, vol. 10013, pp. 172–177. Springer, Cham (2016). doi:10.1007/978-3-319-47440-3_19

Assessing Learning Gains

Jekaterina Rogaten(✉), Bart Rienties, and Denise Whitelock

Open University, Milton Keynes, UK
Jekaterina.rogaten@open.ac.uk

Abstract. Over the last 30 years a range of assessment strategies have been developed aiming to effectively capture students' learning in Higher Education and one such strategy is measuring students' learning gains. The main goal of this study was to examine whether academic performance within modules is a valid proxy for estimating students' learning gains. A total of 17,700 Science and Social Science students in 111 modules at the Open University UK were included in our three-level linear growth-curve model. Results indicated that for students studying in Science disciplines modules, module accounted for 33% of variance in students' initial achievements, and 26% of variance in subsequent learning gains, whereas for students studying in Social Science disciplines modules, module accounted for 6% of variance in initial achievements, and 19% or variance in subsequent learning gains. The importance of the nature of the consistent, high quality assessments in predicting learning gains is discussed.

Keywords: Learning gains · Grades · Assessment · Multilevel modelling · Higher education

1 Introduction

Over the years a variety of assessment strategies have been developed aiming to effectively capture students' learning in Higher Education (HE) [1]. Througout the HE sector universities are using summative assessment, but there is now an increasing number of institutions which are using Computer Based Assessment (CBA) to deliver, monitor, and evaluate students' learning [2–4]. The feedback students receive from CBA is often limited to a grade [2, 5], however there is also formative CBA that is used to inform students and educators of learning progress [6]. Information provided by formative CBA can help to shape learning, and is particularly useful when it is available to learners either before they start work or during the learning process [2, 5, 7, 8].

Given the near universal nature of assessing students' learning in HE, several researchers have used assessment results as proxies for learning gains, which are defined in this article as the change in knowledge, skills, and abilities over time as a result of targeted learning process [9–12]. There are multiple learning gains that students can develop in HE, which are linked to the learning outcomes or learning goals of the course: development of the conceptual understanding of the topic [13]; scientific reasoning and confidence in reasoning skills [14]; scientific writing and reading [15]; critical thinking [16]; problem solving, creativity, analytical ability, technical skills and communication

[17]; moral reasoning [18]; leadership [19]; interest in political and social environment [20]; well-being [21]; and motivation [22]. Measuring such a variety of learning gains is a challenge in itself and a number of methodologies have been used to assess them. The approaches range from pre-post testing using standardised tests to cross-sectional studies using self-reported measures. Assessment of learning gains in knowledge and understanding is no exception and different methods are routinely used.

For example, Hake [13] examined students' learning gains in conceptual understanding of Newtonian mechanics in a sample of 6,542 undergraduate students using standardized tests at the beginning (pre-test) and at the end (post-test) of the course. Similar studies were undertaken by other teams of researchers [23, 24] who also used standardised test in the beginning and end of a semester to capture students' learning gains. These studies reported students making low to moderate learning gains during the limited time of one semester.

A recent literature review by Rogaten and colleagues [25] amongst 51 learning gains studies indicated that with regards to cross-sectional studies, knowledge and understanding along with other learning gains were most often measured with Student Assessment of Learning Gains (SALG) scale [17, 26, 27]. SALG is a self-reported questionnaire that assesses students' perceived level of learning gains. There are also other measures that can be used to reliably assess students' perceptions of learning gains in knowledge and understanding, such as Science Students Skills Inventory (SSSI) [28, 29], and Student Experience in the Research University Survey (SERU-S) [30]. Since these instruments use self-reported measures, these type of studies rely on the premise that students can realistically and adequately appraise their own learning gains, which of course can be disputed [31].

The use of objective tests and pre-post testing to capture students' learning is generally preferred over the use of self-reported measures. Objective tests may capture unbiased learning gains rather than the perceptions of learning gains, and therefore are less reliant on individuals' abilities to properly self-appraise their own learning progress. However, pre-post testing is more resource-intensive in comparison to administration of self-reported surveys at the end of modules, and may become even more cost-intensive if teachers, universities, and governments want to estimate learning gains across various disciplines and number of universities [32].

A potential alternative to the administration of pre-post tests for assessing students' gains in knowledge and understanding is to estimate students' learning gains from course assessments grades. This approach capitalises on the large quantity of student data routinely gathered by every university and, at the same time, offers opportunities to measure learning gains across various disciplines and universities without additional measurement and financial costs. Furthermore, using students' academic performance as a measure of learning progress has other advantages; firstly, it is widely recognized as an appropriate measure of learning, secondly, it is relatively free from self-reported biases, and thirdly, using academic performance allows a direct comparison of research into findings with the results from other studies [33–35].

At the same time, using academic performance scores as proxies for learning might have several limitations, such as a lack of assessment quality (e.g., too easy or too hard, focused on knowledge reproduction rather than critical evaluation) [2, 36], low inter-rater reliability (i.e., two markers give different assessment scores), and/or lack of coherence of assessment difficulty throughout the module (e.g., hard first assessment and easy final assessment; simple first assessment, hard second assessment, easy final assessment) [37, 38]. Therefore, in this article we will ask the following two research questions:

1. To what extent do assessment scores provide a valid, reliable proxy of estimating students' learning gains
2. How much variance in students' learning gains is accounted for by assessments, module characteristics and socio-demographic factors (i.e., gender, ethnicity and prior educational experience)?

In this study, we will use a three-level growth-curve model estimated for 17,700 HE students studying in two distinct disciplines (Science and Social science) in 111 modules at the Open University UK. After a brief review of assessment and feedback literature, we will review how researchers have used assessments as proxies for learning gains.

1.1 Importance of Assessment and Feedback

The majority of HE institutions use assessment and feedback as a driver for and of learning. CBA has a lot of potential applications [4, 39] and benefits are being realized. There are a number of definitions and applications of CBA, but in the context of this study we conceptualize CBA as assessment presented using digital means and submitted electronically. CBA has numerous advantages [40] when compared to other, more traditional types of assessments. The most relevant benefits in distance-learning settings include more authentic interactive assessment options, such as intelligent tutoring [41], authentic virtual labs [42], speed of assessment, automatic feedback [43], and record-keeping. Although CBA is often used for summative assessments to evaluate what students learned, there has been an increase in use of CBA as a formative assessment in a form of online practice quizzes, wikis and peer assessment to provide formative feedback for students [2, 7, 44–46]. Using CBA for summative assessment only provides feedback in a form of a grade once all learning activities are completed [2, 5], whereas using CBA for formative assessment provides information that can help to shape learning, and is particularly useful when it is available to learners either before they start work or during the learning process [6]. As such, CBA is a valuable tool for helping students to regulate their learning processes [2, 5, 7, 8].

A vast body of research has indicated that providing feedback is more important for learning than the assessment of learning [7]. Feedback originates from the field of engineering and information theory with the general assumption that information about the current system's state is used to change the future state. In his meta-study of 800+ meta-studies, Hattie [7] found that the way in which students receive feedback is one of the most powerful factors associated with the enhancement of learning experiences. Hattie and Yates [47] (p. 60) consider feedback as empowering because it enables the learner to "move forward, plot, plan, adjust rethink and exercise self-regulation". For example, Whitelock [48] has argued that feedback is rather restrictive in nature when formative assessment's focus is that of "Assessment for Learning". She suggests that what is required in this context is a concept known as "Advice for Action". This approach does not restrict itself to giving advice after a task has been completed but can also embrace hints given before an assessment task is taken up.

1.2 Measuring and Computing Learning Gains

In the field of learning gains research, only a couple of studies have estimated learning gains from students' academic performance [49, 50] and overall students showed on average a decrease in their grades from the first to the last assessment of a semester/ course. For example, Jensen and colleagues [49] assessed 108 biology course students and used results of 3 interim exams to estimate students' learning gains. Although in their study they focused on how students differed in flipped and non-flipped classrooms in terms of their academic performance across assessments, they reported that over the three unit exams students' performance generally decreased from 81.7% to 75.9% in non-flipped classroom and from 82.4% to 76.3% in flipped classroom. Thus, this decrease was equivalent in both groups of students. Yalaki [50] similarly assessed 168 organic chemistry students using their performance on 2 mid-term examinations and final exam. The goal of this study was to compare whether formative assessments and feedback resulted in better students' attainments in comparison to no formative assessment. They found that performance gradually decreased from the first interim exam to the final examination result (i.e., from 87.3% to 68.8% for group receiving formative assessments and feedback, and from 66% to 61.4% for group receiving no formative assessments). In both of these studies researchers did not examine students' learning gains per se, but rather were interested in group differences in attainment on any one assessment. However, the observed decrease in attainments throughout the semester is contrary to what was found in pre-post test studies using standardized tests [13, 23, 24], where student on average showed an increase in their knowledge and understanding.

In addition to using different means to assess students' learning gains that seem to provide different results, there are a number of ways to compute students' learning gains [9, 11, 13, 51–53]. On the one hand, if one wants to examine the level of knowledge students developed over a course, one would assume that subtracting the beginning of a semester knowledge test score from the end of a semester knowledge test score will produce an accurate level of change/gain in academic achievement. Although this computation of learning gain makes intuitive sense, raw gain as a value of gain is inaccurate due to the difference between scores being less reliable than scores themselves [11], thus, it does not account for random error of measurements between pre-test and post-test scores [9, 10, 52, 54].

Several potential alternatives to raw difference computations have been proposed, such as computation of true gain [11, 12], residual gain [9], normalised gain [13, 55], average normalised gain [51], normalised change [53], ANOVA and ANCOVA on residuals or pre-post test scores [52]. Although these alternatives address the issue of measurement error, all of these methods assume that errors between participants are uncorrelated and, as such, assume that pre-test and post-test observations from one participant are independent from pre-post test observations of another participant. This assumption may not necessarily be true, as students from the same discipline, same class, and/or same university have shared variance due to the similarity of experiences, and this variance is usually overlooked [56]. One way of addressing this limitation is to analyze learning gains within a student as well as between students on a same course. Multilevel growth-curve modeling allows for estimating individual learning trajectories by fitting an overall average course curve and allowing each individual students' curve to depart from the average course curve. Moreover, using multilevel modelling it is possible to estimate what is the variance in students' initial achievements and their subsequent learning gains depending on what module they are enrolled in and whether students' initial achievements and learning gains depend on their individual differences and socio-demographic characteristics.

Several researchers have found that disciplinary differences significantly influence students' learning processes and academic performance. For example, Rienties, and Toetenel [57] found that the way teachers in their respective disciplines designed 151 modules significantly influenced how students were learning in the virtual learning environment, which in turn impacted on student satisfaction and performance. Although course characteristics are important predictors of learning, socio-demographic variables also have been found to play an important role. Thus, some researchers found that there was a gap in attainment in gender with male students being awarded higher final degree classifications than female students [58], whereas in other studies opposite was found i.e., male students were having lower initial academic achievements in comparison to female students, and the gap between males and females increased over time [59]. Ethnicity was also continuously found to be important factor in academic attainment across different levels of education, with white students having higher attainments at all levels of educational system than non-white students [60, 61]. Research also overwhelmingly shows that prior educational attainment is one of the strongest predictors of educational attainment [62, 63], with students who had high academic achievements prior to enrolling into a degree level module are more likely to have high attainments at the degree level.

In light of the challenges facing mass standardized assessments [44, 64] and assumptions on which learning gains computations are based, this study aims to test whether the estimation of a multilevel growth-curve model that accounts for the correlation of errors between participants can be effectively used in predicting students' learning gains from academic performance. As such, the first question this study will address is how much students vary in their initial achievements and their subsequent learning gains in Science and Social Science disciplines? Secondly, taking into account that previous research indicated that there are gender differences in students' achievements and progress (i.e., white students tend to perform better that students from other ethnic backgrounds), and that prior educational experience is a strong predictor of future academic success, this study will also examine whether students' initial achievements and subsequent learning gains depend of student gender, ethnicity and prior educational qualification. Finally, within learning gains research learning gains are traditionally examined in Science students and other disciplines are largely ignored. This study aims to address this gap by estimating multilevel growth-curve models separately for Science and Social Science student samples. It was hypothesized that:

H1: There will be difference in students' learning gains between Science and Social Science disciplines.

H2: There will be an effect of gender, ethnicity and prior educational qualification on students' initial achievements and subsequent learning gains.

2 Method

2.1 Setting and Participants

The Open University UK is a distance-learning institution with an open-entry policy, which is the largest university in the UK. Given that, the OU is open to all people and no formal qualification requirements are present at level 1 modules. Academic performance data for 17,700 undergraduate students from Social Science and from Science faculties was retrieved from an Open University UK database. Social Science student sample comprised of 11,909 students of whom 72% were females and 28% were males with average age of $M = 30.6$, $SD = 9.9$. At the time of registering for the course 43.5% of students had A levels or equivalent qualification, 35.6% had lower that A levels, 15.7% had a HE qualification, 2.4% had postgraduate qualification, and remaining 2.8% had no formal qualification. It is important to note that in majority of UK universities A to C grades at A levels are standards for admission. The majority of students were white (86.8%) followed by black (5%), Asian (3.2%) and mixed and other (5%) ethnic backgrounds.

Science student sample comprised of 5,791 students of whom 58.2% were females and 41.8% were males with average age of $M = 29.8$, $SD = 9.6$. At the time of registering for the course 43.7% of students had A levels or equivalent qualification, 28.8% had lower that A levels, 21.6% had HE qualification, 3.9% had postgraduate qualification, and remaining 1.9% had no formal qualification. Majority of students were white (87.7%) followed by Asian (4.4%), black (3.3%) and mixed or other (4.7%) ethnic backgrounds.

2.2 Measures and Procedure

Ethics was obtained from Open University Human Research Ethics Committee (AMS ref 215140). Academic performance on Tutor Marked Assessments (TMA) was retrieved from the university database for all students enrolled to all modules within Social Science and Science faculties. TMAs usually comprise of tests, essays, reports, portfolios, workbooks, but do not include final examination scores. TMA was suitable for this study as all 111 modules used in the analysis had a minimum of two TMAs and maximum of seven TMAs. TMA grades provided enough longitudinal data for estimating students' learning gains for a period of one semester (i.e., 40 weeks). Academic performance on each TMA for each module was obtained for two cohorts of students who studied in 2013/14 and 2014/15 academic years. In total, TMA results were recorded for 111 modules across two faculties. In case of some missing TMA scores, a multilevel modelling makes automatic adjustments and estimated growth-curves on existing TMA scores and as such, some missing data is acceptable [65, 66].

2.3 Data Analysis

The data was analyzed using a three-level linear growth-curve model estimated in MLWiN software [65, 66]. Identical models were estimated for Social Science modules and Science modules. In the multilevel model, level 1 variable was students' module TMA (repeated measures time variable), level 2 variable was student/participant and level 3 variable was the respective module students were enrolled in. The dependent variable was students' academic performance on each of the TMAs, with the possible minimum score of 0 and possible maximum score of 100. In line with Rasbash and colleagues [65, 66], students' academic performance was centered to the average of the course academic performance, and the time of assessment was centered to the first assessment in order to make intercept and other data parameters more interpretable. The 3-level nested structure is presented in Fig. 1.

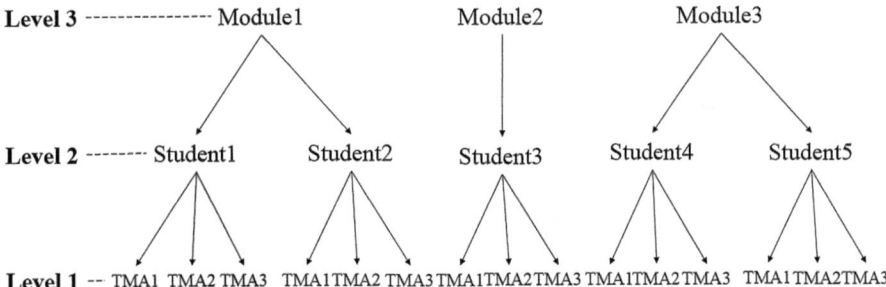

Fig. 1. A three-level data structure with repeated TMA scores at level 1

3 Results

Fitting a multilevel growth curve model to the data as opposite to single-level model (multiple linear regression) significantly improved the fit of the model for both Social Science and Science modules with the likelihood ratio test for Social Science LR = 21929.81, p < 0.001, and for Science students LR = 11537.37, p < 0.001 being significant. Social Science students' academic achievements were on average M = 67.6; SD = 13.7. The results of the growth-curve model estimation showed that module accounted for 6.4% of variance in students' initial achievements, and 18.5% of variance in subsequent learning gains. The student-level intercept-slope correlation was r = 0.138, which indicated that students with initial high achievements and students with initial low achievements progressed at a relatively similar rate. However, a module-level intercept-slope correlation indicated that students in modules with initial low achievements had much higher learning gains than students in modules with initial high achievements(r = −.68). Variance partition coefficient (VPC) showed that in total 3.8% of variance in Social Science students' learning gains could be attributed to the difference between modules, 56% of variance in learning gains could be attributed to individual differences, and 40% of variance was over TMAs within a student i.e., differences between assessments within the module accounted for 40% of total variance. Figure 2 represents students' actual performance, predicted growth-curves for each student, and predicted module growth curves for Social Science.

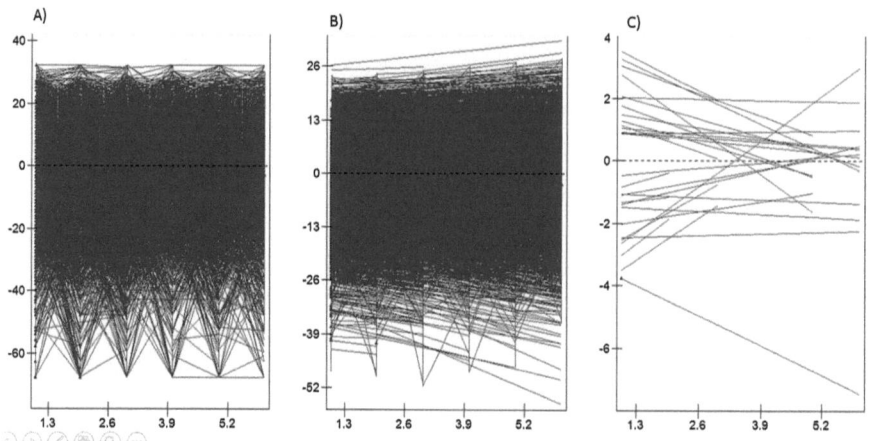

Fig. 2. (A) Trellis plot for student performance on each TMA across all Social Science modules, (B) Predicted student growth-curves across all Social Science modules, and (C) Predicted module growth-curve for each module within Social Science.

Science students' academic achievement was on average M = 65.9; SD = 22.2. The results of the growth curve model estimation showed that 'module' accounted for 33.3% of variance in initial achievements, and 26.4% or variance in subsequent learning gains. The student-level intercept-slope correlation was r = −0.66 indicating that students with initial low achievements showed high learning gains in comparison to students with the high initial achievements. With regards to the module-level intercept-slope correlations, the correlation was r = −0.58 indicating that students in modules with initial low achievements showed higher learning gains than students in modules with initial high achievements. VPC showed that in total 26% of variance in Science students' learning gains could be attributed to the difference between modules, 52% of variance in learning gains could be attributed to individual differences, and only 22% or variance was over TMAs within a student, i.e., differences between assessments within the module accounted for 22% of total variance. Figure 3 represents students' actual performance, predicted growth-curves for each student, and predicted module growth curves for Science.

Comparing Figs. 2 and 3 it is noticeable that in Social Science there was a fanning out in students' predicted growth curves, which indicated that over a period of 40 weeks students with initial high achievements showed an increase in their subsequent achievements, while students with initial low achievements showed a drop in their subsequent achievements. In contrast, this phenomenon was not present for Science students, where students with initial high achievements had lower subsequent achievements, while students who initially had low achievements gradually obtained better grades. On the module level, Social Science modules showed strong fanning in, whereas it was less noticeable in Science modules. This indicated that Social Science students varied much stronger in their assessment results than Science students.

3.1 Influence of Socio-Demographics on Learning Gains

The addition of socio-demographic variables (student level predictors) further improved the fit of the model. Gender explained an additional 3% in Social Science students' initial achievements, with male students showing significantly higher learning gains than female students (Beta = 0.636, $p < 0.01$), and most of this variance was due to the females having lower initial achievements, while there was no gender difference in learning gains for Science students. With regards to ethnicity, white Social Science students showed significantly higher learning gains compared to all other ethnic groups, with the biggest difference being between white and black students (Beta = −7.99, $p < 0.01$), followed by the difference between white and other minority ethnic groups (Beta = −6.68, $p < 0.01$), and between white and Asian students (Beta = −4.66, $p < 0.01$). Overall, ethnicity accounted for an additional 3.4% of variance in Social Science students' subsequent learning gains. White Science students also showed significantly higher learning gains but only in comparison to black students (Beta = −13.07, $p < 0.01$) and Asian students (Beta = −7.31, $p < 0.01$). There were no differences between white and other ethnic groups in their learning gains, and ethnicity only accounted for an additional 2.2% in Science students' subsequent learning gains.

Prior educational qualifications also explained an additional 3% of variance in both Social Science students' learning gains and Science students' learning gains. As one would expect, in Social Science students who started their course having previously obtained a postgraduate qualification showed significantly higher progress than students who only had A levels (Beta = 2.62, $p < 0.05$). Students who had lower than A levels

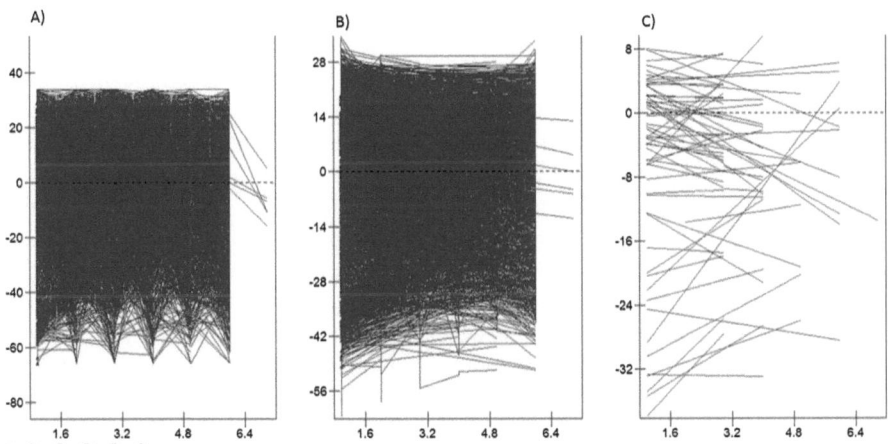

Fig. 3. (A) Trellis plot for student performance on each TMA across all Science modules, (B) Predicted student growth-curves across all Science modules, and (C) Predicted module growth-curve for each module within Science.

achievements or no formal qualification showed significantly lower learning gains than students who had A levels (Beta = −3.11, p < 0.01; Beta = −6.93, p < 0.01 respectively). There were no differences in learning gains between students who had A levels and those who already had an HE qualification. In Science, students who had HE qualification or postgraduate qualification showed significantly higher learning gains than students who only had A levels (Beta = 2.64, p < 0.01; Beta = 8.19, p < 0.01 respectively). Students who had lower than A level qualification or no qualification showed significantly lower learning gains than students who had A levels (Beta = −4.59, p < 0.01; Beta = −9.48, p < 0.01 respectively).

Our results overall supported our two research hypotheses and three-level growth curve models fitted longitudinal assessment data better than single-level models for both Social Science and Science disciplines. In addition, our models explained a significant portion of variance in students' initial achievements and subsequent learning gains. There were also substantial differences between Social Science and Science students in how much variance initial models accounted for, and how much additional variance socio-demographic variables accounted for in students' learning gains.

4 Discussion

The first aim of this research was to examine whether three-level growth-curve modelling on assessment scores was a better alternative to single-level models in capturing students' learning gains from module assessment data. The second aim of this project was to examine whether socio-demographic factors had any effect on students' initial achievements and subsequent learning gins. The third aim was to examine whether there was difference between multilevel models estimated for Social Science students and Science students.

The results overwhelmingly supported the superiority of multilevel growth-curve model for estimating students' learning gains in both Social Science and Science. Overall, the three-level growth-curve model for Science students accounted for more variance in learning gains than the identical model for Social Science. As such, the basic model explained variance in Science students' learning gains better than it did for Social Science students. Despite these differences, multilevel modelling was superior to single level models and as such, a more accurate method for estimating students' learning progress. The advantage of multilevel models is in that simple models are not able to detect differences between modules when looking at discipline level performance, whereas multilevel modelling accounts for those differences. This has important implications for assessing students' learning gains on an institutional level. Furthermore, this provides important policy implications when comparing learning gains across different institutions and faculties, as is currently the intention by the Teaching Excellence Framework in the UK and policy initiatives of standardized testing in the US.

In particular, an interesting finding was that Social Science students tended not to differ in progress they made regardless of their initial achievements, while amongst Science students initially low achievers were able to obtain higher learning gains over time. This finding could be due to variety of factors, but one possible explanation is that

Science students' performance is more stable throughout a semester than Social Science students' performance. This is partly due to the assessments used in different disciplines to test students' knowledge. In Science, knowledge tends to be assessed using tests, workbooks and examinations, whereas in Social Science assessments are much more diverse, including essays, reports, portfolios and reviews. VPCs for each discipline further supported this interpretation, with results showing that while in Science 22% of variance in performance was across different TMAs, for Social Science students this variance was almost doubled reaching 40%. Thus, in case of Social Sciences, students may take longer to learn how to present/show their understanding and knowledge i.e., the ability to write a good essay for the first assessment does not guarantee that a student will be able to write a good report or review article for the next assessment, and hence there is greater variability in TMA scores. As such, it may be harder for Social Science students with low initial achievements to show learning gains that are higher than those with initial high achievements, despite the fact that low achievers have more room for improvement than high achievers due to obvious ceiling effects. Different patterns were observed amongst Science students, where initial low achievers outperformed initial high achievers on their rate of progress i.e., learning gains indicating that there could be a potential ceiling effect.

Another important finding of this study is that there were several modules where no learning gains or even negative ones were observed, while several modules did show large positive learning gains. Negative learning gains were previously reported in the literature and were only observed when they were estimated on students' assessments [49, 50]. However, negative learning gains that were mainly observed among students and modules with high initial achievements does not automatically imply that students are losing knowledge or understanding per se. However, it does highlight the complexity of factors that have to be taken into account when using students' academic performance as a proxy for learning gains. These factors include assessment difficulty, consistency of assessment rubrics over time, and learning design [57].

Overall, our multi-level growth methodology proposed in this article starts to assess the 'distance travelled' by a student in terms of their learning over the duration of a course of study. The assessment scores that could be derived from these CBAs could facilitate a deeper dive into a more exact description of student achievement and learning. In other words, it opens the possibility of retrieving information automatically about which questions are answered well, and which skills and competencies the students are acquiring. Well contrasted CBAs can support the measurement of learning gains at the level of competences which are of interest to academics and employers alike.

With regards to the effects of demographic factors (i.e., gender, ethnicity, and prior academic qualifications) on learning gains accounted for additional 6.4% in Social Science students' learning gains, which is larger than additional variance accounted for in Science students' learning gains (5%). Out of all socio-demographic variables, the strongest predictor for learning gains was prior educational experience/qualification with students who had A levels and above showing significantly higher learning gains than those who had below A levels or no qualification. This was closely followed by ethnicity, with white students showing highest learning gains in comparison to other ethnic groups. These findings highlight that these differences can possibly lie in different experience of HE, where white students with minimum good grades on A levels form a majority of HE students in UK [61]. However, current government plans to increase diversity of students from different ethnic and socio-economic backgrounds calls for more research into how "non-traditional" students are progressing in HE. Attracting students who are "disadvantaged" and who may be possibly the first generation in their family to attend HE also implies that universities should be actively helping those students to develop basic study skills and assist them in learning how to study effectively at an HE level [67]. Despite the fact that the initial starting point of a student might be below average, the provision of adequate support is likely to decrease the gap in students' learning gains over time between traditional and non-traditional students.

Although the results of our study are important for understanding students' learning gains in HE, this research has number of limitations that should be taken into account when interpreting and generalizing our findings. Firstly, performance data was only collected from samples of learners who were enrolled in Science and Social Science modules at one distance learning university. Because those students only usually take one module during a semester, their pace of learning could be different compared to students who are in full-time, face-to-face education and usually take four modules in one semester. Secondly, learning gains were estimated only for a limited time of a semester of 40 weeks, and as such it is possible that learning gains and observed effects of socio-demographic factors could be different when multilevel models are estimated across semesters or years. Thus, the same student could show different patterns of progress across semesters and across years. As such, future research should aim to collect longitudinal data across several years of study for full-time and part-time modules across different institutions in order to validate the generalization of our findings. Thirdly, only academic performance of students and socio-demographic data was used to estimate and explain variance in students' learning gains. Taking into account increasing use of Virtual Learning Environment (VLE) across universities, future research should aim to collect data on students' actual learning behavior, e.g., discussion forums participation, time spent going through study materials, access to additional materials and library resources. By examining differences in patterns of VLE behavior for students who make high or low learning gains, more insight could be obtained about the underlying reasons for different learning gains between students. Finally, in this study grades from the module assessments were used to estimate learning gains and detailed examination of the nature of assessments and difficulty of assessments was not taken into account. As such, future research should look more in-depth into different assessment formats and approaches used across various modules and to control for the assessment difficulty when estimating students' learning gains. Despite these limitations, the methodology of using multi-level growth modelling to understand to what extent students are making different learning gains over time seems very appropriate. In particular, our three-level linear growth-curve model can highlight differences and inconsistencies in assessment practices within and across modules, and help teachers and senior management to ensure a consistent quality of assessment provision.

References

1. Brown, G.A., Bull, J., Pendlebury, M.: Assessing Student Learning in Higher Education. Routledge, London (2013)
2. Boud, D., Falchikov, N.: Aligning assessment with long-term learning. Assess. Eval. High. Educ. **31**, 399–413 (2006)
3. Nicol, D.: From monologue to dialogue: improving written feedback processes in mass higher education. Assess. Eval. High. Educ. **35**, 501–517 (2010)
4. Tempelaar, D.T., Rienties, B., Giesbers, B.: In search for the most informative data for feedback generation: learning analytics in a data-rich context. Comput. Hum. Behav. **47**, 157–167 (2015)
5. Whitelock, D., Richardson, J., Field, D., Van Labeke, N., Pulman, S.: Designing and testing visual representations of draft essays for higher education students. In: Presented at the Paper presented at the Learning Analytics Knowledge Conference, Indianapolis (2014)
6. Segers, M., Dochy, F., Cascallar, E. (eds.): Optimising New Modes of Assessment: In Search of Qualities and Standards. Kluwer Academic Publishers, Dordrecht (2003)
7. Hattie, J.: Visible Learning: A Synthesis of Over 800 Meta-Analyses Relating to Achievement. Routledge, London (2008)
8. Lehmann, T., Hähnlein, I., Ifenthaler, D.: Cognitive, metacognitive and motivational perspectives on preflection in self-regulated online learning. Comput. Hum. Behav. **32**, 313–323 (2014)
9. Cronbach, L.J., Furby, L.: How we should measure "change": or should we? Psychol. Bull. **74**, 68–80 (1970)
10. Linn, R.L., Slinde, J.A.: The determination of the significance of change between pre- and posttesting periods. Rev. Educ. Res. **47**, 121–150 (1977)
11. Lord, F.M.: The measurement of growth. ETS Res. Bull. Ser. **1956**, i-22 (1956)
12. Lord, F.M.: Further problems in the measurement of growth. Educ. Psychol. Meas. **18**, 437–451 (1958)
13. Hake, R.R.: Interactive-engagement versus traditional methods: a six-thousand-student survey of mechanics test data for introductory physics courses. Am. J. Phys. **66**, 64–74 (1998)
14. Beck, C.W., Blumer, L.S.: Inquiry-based ecology laboratory courses improve student confidence and scientific reasoning skills. Ecosphere, **3**, UNSP 112 (2012)
15. Coil, D., Wenderoth, M.P., Cunningham, M., Dirks, C.: Teaching the process of science: faculty perceptions and an effective methodology. CBE-Life Sci. Educ. **9**, 524–535 (2010)
16. Mortensen, C.J., Nicholson, A.M.: The flipped classroom stimulates greater learning and is a modern 21st century approach to teaching today's undergraduates. J. Anim. Sci. **93**, 3722–3731 (2015)
17. Gill, T.G., Mullarkey, M.T.: Taking a case method capstone course online: a comparative case study. J. Inf. Technol. Educ.-Res. **14**, 189–218 (2015)
18. Mayhew, M.J.: A multilevel examination of the influence of institutional type on the moral reasoning development of first-year students. J. High. Educ. **83**, 367–388 (2012)
19. Riutta, S., Teodorescu, D.: Leadership development on a diverse campus. J. Coll. Stud. Dev. **55**, 830–836 (2014)
20. Pascarella, E.T., Salisbury, M.H., Martin, G.L., Blaich, C.: Some complexities in the effects of diversity experiences on orientation toward social/political activism and political views in the first year of college. J. High. Educ. **83**, 467–496 (2012)
21. Bowman, N.A.: The development of psychological well-being among first-year college students. J. Coll. Stud. Dev. **51**, 180–200 (2010)

22. Trolian, T.L., Jach, E.A., Hanson, J.M., Pascarella, E.T.: Influencing academic motivation: the effects of student-faculty interaction. J. Coll. Stud. Dev. (in press)
23. Andrews, T.M., Leonard, M.J., Colgrove, C.A., Kalinowski, S.T.: Active learning not associated with student learning in a random sample of college biology courses. CBE-Life Sci. Educ. **10**, 394–405 (2011)
24. Cahill, M.J., Hynes, K.M., Trousil, R., Brooks, L.A., McDaniel, M.A., Repice, M., Zhao, J., Frey, R.F.: Multiyear, multi-instructor evaluation of a large-class interactive-engagement curriculum. Phys. Rev. Spec. Top.-Phys. Educ. Res. **10**, 020101–020119 (2014)
25. Rogaten, J., Rienties, B, Sharpe, R., Cross, S., Whitelock, D., Lygo-Baker, S., Littlejohn, A.: Reviewing affective, behavioural and cognitive learning gains in higher education. (Submitted)
26. Anderson, A.: An assessment of the perception of learning gains of freshmen students in an introductory course in nutrition and food science. J. Food Sci. Educ. **5**, 25–30 (2006)
27. Ojennus, D.D.: Assessment of learning gains in a flipped biochemistry classroom. Biochem. Mol. Biol. Educ. **44**, 20–27 (2016)
28. Hodgson, Y., Varsavsky, C., Matthews, K.E.: Assessment and teaching of science skills: whole of programme perceptions of graduating students. Assess. Eval. High. Educ. **39**, 515–530 (2014)
29. Varsavsky, C., Matthews, K.E., Hodgson, Y.: Perceptions of science graduating students on their learning gains. Int. J. Sci. Educ. **36**, 929–951 (2014)
30. Douglass, J.A., Thomson, G., Zhao, C.-M.: The learning outcomes race: the value of self-reported gains in large research universities. High. Educ. **64**, 317–335 (2012)
31. Richardson, J.T.E.: The attainment of white and ethnic minority students in distance education. Assess. Eval. High. Educ. **37**, 393–408 (2012)
32. Farrington, D.P.: Longitudinal research strategies: advantages, problems, and prospects. J. Am. Acad. Child Adolesc. Psychiatry **30**, 369–374 (1991)
33. Anaya, G.: College impact on student learning: comparing the use of self-reported gains, standardized test scores, and college grades. Res. High. Educ. **40**, 499–526 (1999)
34. Bowman, N.A.: Can 1st-year college students accurately report their learning and development? Am. Educ. Res. J. **47**, 466–496 (2010)
35. Gonyea, R.M.: Self-reported data in institutional research: review and recommendations. New Dir. Institutional Res. **2005**, 73–89 (2005)
36. Biggs, J., Tang, C.: Teaching For Quality Learning at University. McGraw-Hill Education, UK (2011)
37. Carless, D., Salter, D., Yang, M., Lam, J.: Developing sustainable feedback practices. Stud. High. Educ. **36**, 395–407 (2011)
38. Carless, D.: Trust, distrust and their impact on assessment reform. Assess. Eval. High. Educ. **34**, 79–89 (2009)
39. Ras, E., Whitelock, D., Kalz, M.: The promise and potential of e-Assessment for learning. In: Reimann, P., Bull, S., Kickmeier-Rust, M.D., Vatrapu, R., Wasson, B. (eds.) Measuring and Visualising Learning in the Information-Rich Classroom, pp. 21–40. Routledge, London (2016)
40. Terzis, V., Economides, A.A.: The acceptance and use of computer based assessment. Comput. Educ. **56**, 1032–1044 (2011)
41. Koedinger, K.R., Booth, J.L., Klahr, D.: Instructional complexity and the science to constrain it. Science **342**, 935–937 (2013)
42. Scherer, R., Meßinger-Koppelt, J., Tiemann, R.: Developing a computer-based assessment of complex problem solving in Chemistry. Int. J. STEM Educ. **1**, 2 (2014)

43. Whitelock, D.: Maximising student success with automatic formative feedback for both teachers and students. In: Ras, E., Joosten-ten Brinke, D. (eds.) CAA 2015. CCIS, vol. 571, pp. 142–148. Springer, Cham (2015). doi:10.1007/978-3-319-27704-2_14
44. Price, M., O'Donovan, B., Rust, C., Carroll, J.: Assessment standards: a manifesto for change. Brookes eJournal Learn. Teach. **2**, 1–2 (2008)
45. van Zundert, M., Sluijsmans, D., van Merriënboer, J.: Effective peer assessment processes: Research findings and future directions. Learn. Instr. **20**, 270–279 (2010)
46. Whitelock, D., Gilbert, L., Gale, V.: e-Assessment tales: what types of literature are informing day-to-day practice? Int. J. E-Assess. 3, (2013)
47. Hattie, J., Yates, G.C.R.: Visible Learning and the Science of How We Learn. Routledge, New York (2013)
48. Whitelock, D.: Activating assessment for learning: are we on the way with web 2.0? In: Lee, M.J.W., McLoughlin, C. (eds.) Web 2.0-Based e-Learning: Applying Social Informatics for Tertiary Teaching. IGI Global (2010)
49. Jensen, J.L., Kummer, T.A., Godoy, P.D.D.M.: Improvements from a flipped classroom may simply be the fruits of active learning. CBE-Life Sci. Educ. **14**, 1–12 (2015)
50. Yalaki, Y.: Simple formative assessment, high learning gains in college general chemistry. Egitim Arastirmalari-Eurasian J. Educ. Res. **10**, 223–243 (2010)
51. Bao, L.: Theoretical comparisons of average normalized gain calculations. Am. J. Phys. **74**, 917–922 (2006)
52. Dimitrov, D.M., Rumrill Jr., P.D.: Pretest-posttest designs and measurement of change. Work **20**, 159–165 (2003)
53. Marx, J.D., Cummings, K.: Normalized change. Am. J. Phys. **75**, 87–91 (2007)
54. Pike, G.R.: Lies, damn lies, and statistics revisited a comparison of three methods of representing change. Res. High. Educ. **33**, 71–84 (1992)
55. Hake, R.R.: Lessons From the Physics-Education Reform Effort (2001). arXiv:physics/0106087
56. Snijders, T.A.B., Bosker, R.: Multilevel Analysis: An Introduction to Basic and Advanced Multilevel Modeling. Sage Publications Ltd., London (2011)
57. Rienties, B., Toetenel, L.: The impact of learning design on student behaviour, satisfaction and performance: a cross-institutional comparison across 151 modules. Comput. Hum. Behav. **60**, 333–341 (2016)
58. Mellanby, J., Martin, M., O'Doherty, J.: The "gender gap" in final examination results at Oxford University. Br. J. Psychol. **91**, 377–390 (2000)
59. Conger, D., Long, M.C.: Why are men falling behind? Gender gaps in college performance and persistence. Ann. Am. Acad. Pol. Soc. Sci. **627**, 184–214 (2010)
60. Kao, G., Thompson, J.S.: Racial and ethnic stratification in educational achievement and attainment. Ann. Rev. Sociol. **29**, 417–442 (2003)
61. Richardson, J.T.E.: The under-attainment of ethnic minority students in UK higher education: what we know and what we don't know. J. Furth. High. Educ. **39**, 278–291 (2015)
62. Plant, E.A., Ericsson, K.A., Hill, L., Asberg, K.: Why study time does not predict grade point average across college students: implications of deliberate practice for academic performance. Contemp. Educ. Psychol. **30**, 96–116 (2005)
63. Rogaten, J., Moneta, G.B.: Positive and negative structures and processes underlying academic performance: a chained mediation model. J. Happiness Stud. 1–25 (2016)
64. Horn, C.: Standardized assessments and the flow of students into the college admission pool. Educ. Policy. **19**, 331–348 (2005)
65. Rasbash, J., Charlton, C., Browne, W.J., Healy, M., Cameron, B.: MLwiN Version 2.02. University of Bristol, Centre for Multilevel Modelling (2005)

66. Rasbash, J., Steele, F., Browne, W.J., Goldstein, H.: A User's Guide to MLwiN. Cent. Multilevel Model. Univ. Bristol, 296 (2009)
67. Sanders, J., Rose-Adams, J.: Black and minority ethnic student attainment: a survey of research and exploration of the importance of teacher and student expectations. Widening Particip. Lifelong Learn. **16**, 5–27 (2014)

Requirements for E-testing Services in the AfgREN Cloud-Based E-learning System

Salim Saay[1(✉)], Mart Laanpere[2], and Alex Norta[3]

[1] Computer Science Faculty, Kabul University, Kabul, Afghanistan
saay@tlu.ee
[2] Institute of Informatics, Tallinn University, Narva Road 25, 10120 Tallinn, Estonia
mart.laanpere@tlu.ee
[3] Department of Informatics, Tallinn University of Technology, Akadeemia tee 15A, 12816 Tallinn, Estonia
alex.norta.phd@ieee.org

Abstract. National Research and Education Networks (NRENs) that already exist in many countries are a promising infrastructure for nationwide e-assessments. However, researchers do not focus sufficiently on NREN based e-assessment architectures. Many e-assessment architectures based on public cloud computing already exist that have interoperability, security and adaptability problems. Stand-alone and web-based learning management systems (LMSs) have also limitations of adaptability for the nationwide e-assessment. In this paper, we investigate how to improve the interoperability, security and adaptability of e-assessment systems on the nationwide level. The Afghanistan research and education network (AfgREN) is considered as a case in this research. A requirements bazaar is used for gathering data about functional and non-functional requirements. We use design science as a research method and we analyze the relevant aspects of NREN-based e-assessment architectures by using the architecture tradeoff analysis method (ATAM).

Keywords: E-assessment · University · Entrance exam · AfgREN · Architecture · Cloud computing

1 Introduction

Educational organizations introduced e-assessments more than 16 years ago [1]. Many stand-alone and web-based learning management systems (LMS) exist and comprise elements of e-assessments. For example, Moodle [2] supports different e-assessments including students assignment, submitting of assignments and feedback of teachers. Still, these systems cannot be used for nationwide assessments. Modern e-learning and e-testing systems that include service-oriented architectures (SOA), use cloud computing [1] as the most recent technology and improvement in the area of computing infrastructure. Instructional Management Systems Global Learning Consortium (IMS GLC, or IMS) [3] is one of the learning management systems that works to ensure reusability, portability, platform independence, and longevity for e-testing. It considers the stimulating production of high-quality testing content by making it open to big communities of designers. It also focuses on the amplification of aspects of product analysis and design by using ready-to-use templates in the form of data models. IMS considers the increasing of the overall market size as products can be specialized and, diversified and thus, new audiences can be found[1]. Question and Test Interoperability (QTI) [4] is one of the components of IMS that was released in 2009. Question & Test Interoperability (QTI) [4] is one of the most popular standards for e-testing. QTI is a web-based application that has many different versions and different platforms, e.g. a mobile platform. QTI also supports different types of questions, including multiple choice and text entry. QTI has three sub-components, including (1) assessment administration, registering, and rostering (2) Iteam bank and test bank (3) assessment results and outcomes analytics. An *authoring system, test bank, an assessment delivery system are the main* components of the QTI, Data exchange between authoring tools, item banks, test constructional tools, learning platforms and assessment delivery systems. It gave opportunity that author can manage the system. This system can be a good example for the standardization of e-testing. However, it can never cover all requirements for a nationwide e-testing. İnteroperability, integrity and security are always a concern in e-assessment systems. Many countries have specific requirements and some specific types of assessments for which a national e-assessment architecture is needed. The university entrance exam in Afghanistan is an example of nationwide assessment, where around 250,000 students from all schools of the country participate.

Cloud computing is a promising infrastructure for implementing e-assessments to provide interoperability between different learning-management systems (LMS) [5]. Still, the resource scheduling problem [6] in a public cloud computing infrastructure is an open question. Limitations of data security, high performance and interoperability exist in public cloud computing [7]. Therefore, in this research we consider the National Research and Education Networks (NRENs) as a private cloud for educational organizations. We propose an e-assessment system based on the NREN architecture. The university entrance exam in Afghanistan is used as a case in this research.

National Research and Education Networks (NRENs) are for sharing resources between member organizations. NRENs play an important role in e-learning activities as they enable cross-organizational communication [8]. Currently, there are around 120 NRENs operating worldwide [9] that provide collaboration and sharing resources across national organization and between different countries. Since NRENs exist in many countries and provide high-speed communication between educational and research organizations, research needs to focus on e-testing architecture based on the NREN infrastructure to facilitate the interoperability of the existing LMS systems. Our literature review [1, 5, 10–13] demonstrates that the current e-assessment systems do not cover the full requirements for the national level.

Our goal in this research is to explore the functional and non-functional requirements of e-testing for nationwide e-assessment systems on a cross-organizational level to facilitate reusability and sharing resources. We attempt to improve the interoperability, security, performance, and collaboration between relevant organizations by the use of

---

[1] https://www.imsglobal.org/question/qtiv2p2/imsqti_v2p2_oview.html#h.z12cx3m4z4p5.

the NREN architecture [14]. By considering the above problem and the NREN infrastructure, we pose the following questions: How to design a nationwide e-testing system to improve interoperability, security and performance? To establish a separation of concerns, we deduce the following sub questions. What functional and nonfunctional requirements exist for the e-assessment process based on the AfgREN infrastructure? What current processes exist for knowledge assessment of students in Afghanistan? What are the main benefits of a nationwide e-testing system?

In this research, we analyze the e-testing systems, investigate the requirements for the implementation of the e-testing on a national level, and we design a suitable e-testing architecture based on the National Research and Education Network (NREN). The Afghanistan National Research and Education Network (AfgREN) is considered as a case for this research. We use the design science research method [15] and requirements bazaar [16] for functional and non-functional requirements. We design the e-testing architecture based on the Model Driven Engineering (MDE) method [17] and we evaluate the proposed architecture based on the Architecture Tradeoff Analyses Method (ATAM) [18]. The paper is organized as follows. Section 2 discusses e-testing services and the analysis of current standards in Afghanistan context. Section 3 explores the research design and current process. In Sect. 4 we explore the proposed e-testing architecture, and Sect. 5 concludes the paper together with some suggestions for future work.

2 E-testing Services: Analysis of Standards and Implementation Models

Computer Based Training (CBT) systems in the educational domain became famous after 1998, when a video-based testing of conflict resolution skills was developed [12]. Computer based testing was developed for universities and the process of assessment is often implemented by means of automated testing. Many computer-based applications for testing already exist, including Cisco Networking Academy assessment system [19] and Microsoft assessment system [20] that works worldwide, but they are not for a nationwide e-testing, furthermore various e-learning software exists, mainly termed learning management systems (LMSs) that does not support e-testing [10] due to its complexity. The design of the e-testing system started in 1999 to help educational organizations to administer exams for a large number of students. Miguel et al. [11] proposes an innovative approach for modelling trustworthiness in the context of secure learning assessment in on-line collaborative learning groups, but the network part and data mining remain an open question. The online learning with adaptive testing is a system where questions are stored in a database, and students can learn and pass the test based on web services.

There are many other e-assessment modular architectures that compare based on some criteria such as the authenticity of e-assessments, including availability, integrity, authentication, confidentiality and non-repetition [11]. Gusev and Armenski [1] analyzed different e-assessment architectures and then propose e-assessment on a cloud based solution. Still, cloud-based architectures have many security, ownership- and interoperability concerns [19]. The public cloud is not fully under the control of the owner, and thus, the ownership of resources is an issue. While e-testing needs a secure and controllable infrastructure, the public cloud is not as secure, reliable, interoperable and flexible as it should be.

Sierra et al. [10] propose the integration of new module with LMS to support assessments. Arif et al. [13] propose an integrated e-learning and e-testing system. They consider a five-layer client-server based architecture. The layers of the e-learning architecture include the interface layer, the resource layer, the multi-agent layer, the database controller layer and the database layer. The architectural layers of e-testing include the human user interface layer, the user and admin resource agent, the functional description layer, the implementation layer and the database layer. The human user interface layer is an interface through which the user and the admin can log in to the system. The user can select a form of questions and pass different types of exams. The layer of the user and admin provide a communication facility between students and employers which, the students can post their profiles and the employer can post their vacancies for employees. They use the intelligent agent so that the progress and performance of students can also be checked. Functional description layer is another part of this architecture that operates as a controller of questions bank.

Hajjej et al. [6] propose a new approach to the generic and personalized e-assessment process. They use existing learning management systems and propose the adaption of these systems to cloud computing. However, they do not sufficiently analyzed e-assessment using cloud computing. In Sect. 2.1, we explore the LMS and online learning platforms that are used in Afghanistan research and education network (AfgREN) domain. Based on the design science research method that design of architecture need to be rigor and relevance to environment, thus AfgREN domain considered as relevance environment in this research.

2.1 Afghanistan Research and Education Network (AfgREN)

The AfgREN was established at the public universities, thus, there is a good chance for implementation of e-learning and e-testing system. Public internet connection for Kabul University as the largest university in Afghanistan was established in 2006, but there was no NREN structured. E-learning projects, for example Afghans Next Generation e-Learning (ANGeL) [21], Cisco Networking Academy Program [19] and e-Campus Afghanistan[2] was the projects that implemented in public universities but low internet bandwidth, an insufficient infrastructure was the main problem for them. ANGeL was a project that started in 2006 and ended in 2010. The portal of ANGeL had learning materials, course management tools, and also e-assessment tools. However, the e-assessment tools were never used [21].

The Cisco Networking Academy Program [19] is another good example of using computer based learning and e-testing. The Cisco system e-learning platform enabled the students to register online, take quizzes and online exams. Students could view their grade books and marks online and access the course material online regardless of their

---

[2] http://www.ecampus-afghanistan.org/.

physical location. This system is a real example of a successful e-learning platform at a governmental institution in Afghanistan.

e-Campus Afghanistan (see Footnote 2) is another project initiated in coordination with Afghanistan nine major universities. e-Campus Afghanistan is based on a popular e-learning platform Modular Object-Oriented Dynamic Learning (Moodle) [22] environment. Using the Moodle platform in this programme provides a facility to put courses online. The programme also enables students and lecturers to have interactive communication and discuss a topic in an online forum, but still it does not have any e-testing system.

Currently the ministry of higher education (MoHE) of Afghanistan have a computer based, university entrance exam system called cancour that has three databases, including the student registration database, the question bank databased and the result analysis database. Candidates for university entrance exam introduce by the ministry of education (MoE) but the record of students transfer paper based to the MoHE, while ministry of education also have a management information system (MIS) that can transfer the record of students to cancour system through a computer network.

The cancour exam is one of the most important exams, which around 300,000 candidates compete for only around 60,000 seats that are available in public universities. The computer-based system of cancour is a web-based application and can be used from anywhere and anytime, but at the moment it works only in a local area network (LAN) of the examination office. The system has the method of preventing repeated questions, generating answer keys and saving questions by subject and class matter. It means the system is designed with all concepts of new technology, but the administrators do not use the proper system with all its components, because of network security concern. Furthermore, the MoE also have an examination system for technical institutes, but no collaboration and interoperability are between the systems. Since the MoE and the MoHE both are very serious about security of the exam, even the cancour system is disconnected from the public network. By considering the AfgREN that already exists in Afghanistan, we propose in this research an adaption of the cancour examination management systems to the AfgREN architecture [14].

The AfgREN is for connecting public and private universities, higher education institutions, colleges, teaching hospitals, libraries, research and scientific centers to the global research community. Currently, the AfgREN provides some public services such as access to e-mail, video conferencing and web browsing. This does not meet the needs of the research community. Now the AfgREN is a good infrastructure for the implementation of e-learning and e-testing. As mentioned above, several e-learning projects such ANGeL, eCampus, and etc., were not successful and sustainable. The lack of access to high- speed network and sustainable infrastructure were one of the main reasons for not being successful. The AfgREN is a private cloud for Afghanistan educational organizations that can provide software as a services (SaaS), infrastructure as a service (IaaS) and platform as a services (PaaS). By using the AgREN as an infrastructure of e-learning and e-testing, Afghanistan universities can use their own hardware and software resources to achieve cloud services. This model mostly uses virtualization, and we will use this model for maintaining and backing up the universities resources, because in the private cloud, all the internal servers are virtualized and only the employees' of universities can access the private cloud.

In this model, we can put the university resources including question bank for e-testing in the AfgREN data center where it migrates and maintains the resources. The AfgREN can operate solely for the educational organizations, either behind an internal firewall operated by a third party for the exclusive use of AfgREN members. It will be managed by the employees of the AfgREN network operation center. We can virtualize all the datacenters of all universities, and then migrate to the AfgREN central data center.

3 Research Design

The design science research method is used in this research. Design science research is applicable in many different fields, including architecture, engineering, education, psychology and fine arts [15]. The design science research method is applied to design the architecture of an e-testing in a relevant and rigorous way. In recent years, design science research has become a popular research method in the field of information systems. Researchers can also use design science for designing and evaluating sociotechnical artifacts that can be either real artifacts or designs of theories [23]. E-learning and e-testing is a sociotechnical system that comprises more than one technical system. A sociotechnical system in an e-learning and e-testing context is an information technology system that has an ongoing social process implementation, usually articulation work that is needed for IT activities. It has a complex relationship topology between several system parts and the application context. It considers the context of IT and the social use [24]. Therefore, we consider design science research as the most suitable method for the architectural design in this research.

We use Requirements Bazaar[3] for collecting functional and nonfunctional requirements of an e-testing system. It works for communication and negotiation between the software developer and the customer. In this research, we use Requirement Bazaar to communicate with relevant people via this portal. Requirements Bazaar is an open, web-based portal for voting on ideas and requirements [16]. Anyone who wants vote on any idea or requirement can browse the portal and vote without any restriction. Sections 3.1 and 3.2 provides an overview of the functional and nonfunctional requirements of the e-testing system.

Model driven engineering (MDE) is used for design of the architecture. MDE is a domain- specific modeling language to formalize the application structure, behavior and the requirements of systems in a specific domain. It also facilitates analyzing specific aspects of models and imposes domain performance checking to detect errors before commencing a development life cycle [25]. A model is an abstraction of a system that is under study or targeted for application-system development in the future [26]. MDE is an approach that focus in modeling technology, it use for conceptual model of software artifact and documentation. There is a subset mode of MDE called model driven development (MDD) that use for software automatic code building [27].

---

[3] http://requirements-bazaar.org/.

We use the ATAM [18] for evaluating and analyzing in a qualitative and empirical way the e-learning and e-testing architecture with the related functional and non-functional requirements. ATAM is mostly used for exploring the use of best-practice architectural styles for quality attributes of architecture and for the evaluation of existing systems. ATAM also helps to modify an architecture or integration work with new systems. The purpose of ATAM is to identify risks that emerge because of architectural design decisions and to mitigate those risks in future efforts for modeling and prototypes. Our goal using of ATAM in this research is to elicit the communication factors that affect the quality of e-learning architectures and to consider the factors for reference in NREN e-testing architecture. We used the e-testing as a scenario of NREN e-learning architecture [14] and we used the result of that ATAM workshop in this paper.

3.1 Functional Requirements

The system should ensure that questions get entered to the database system by university professors only once and the system must define where and how many times a question has been used. This database should be able to generate a booklet of examination questions with every booklet having a unique form number for a specific testing. For better prevention of duplicating questions, the once used questions must be archived in the system so that they would not be used again in another booklet. The system should display various reports for the used and removed numbers of questions and booklets in the system. With a specific booklet for each exam, when the exam starts, the system must generate an answer key for each booklet that should be highly protected and encrypted.

The system will also convince the public/stakeholders and all other concerned individuals of their secrecy. There will be different categories of questions (Easy, Medium, and Difficult) and the system will provide a facility for selecting the level of questions in all these categories for the administrators while making each booklet. There will be also such an option that during a certain time a large number of questions is provided, e.g. 10,000 questions for each subject, and then open access for candidates is given. The system should shuffle all the questions, and after selecting any question to a booklet, this question will be marked as used and cannot be used again. The system will be able to archive each form or booklet to generate its answer key easily. While adding or entering questions to the system, the system should also store the answer for each question. The system have a linkage with the post-examination process. The system must have strong security policy to be accessed only by authorized members. The profiles of candidates are needed to specify eligible candidates. The assessment system must be available through a secure web interface and the MoE should also have access to the registration module of the system to transfer the candidates' profile from the MoE Management Information System (MIS) to the university entrance exam system.

3.2 Nonfunctional Requirements

Modifiability: a NREN e-testing architecture changes and adapts quickly and cost-effectively without a restriction to extensibility and the restructuring of software and hardware updates. Additionally, it harmonizes inter-organizational heterogeneous system environments.

Interoperability: a NREN e-testing architecture must interoperate at runtime with other e-testing systems including cloud-based and other traditional e-testing architectures. It must interoperate without any restricted access or implementation.

integradablity: a NREN e-testing architecture comprises separately developed and integrated components for which the interface protocols between the components must match. Hence, integribility between the components of the NREN e-testing architecture must be ensured. ATAM workshop in [28] specify the system requirements for NREN e-testing architecture that are discernible during runtime, because their effectiveness is investigable during the setup and enactment of collaboration configurations.

Scalability: refers to the ability of the NREN e-testing architecture to combine more than two collaborating parties into one configuration.

Performance: means that the computational and communicational strain is low in collaboration configurations for the setup and enactment phase. Hence, it is important to ensure that all phases of a collaboration are carried out within a desirable response time and without an exponential need for computing power.

Security: is the protection of systems from theft or damage to hardware, software, and the information on the former, as well as from disruption or misdirection of the services they provide.

3.3 Process Model

While a computer based assessment system already exists in Afghanistan, but information exchange between the MoHE and the MoE are still paper-based. Student profiles exist in the MoE, MoHE needs those profiles for the university entrance exam, and getting the profiles on paper based from the MoE means that the MoHE will spend months to enter the data on exam candidate's profiles. Numerous mistakes happen in this duplication of data entry. The steps in the process are the following:

A web based application of university entrance exam system already exist.
the system has been developed based on SQL server
Questions are collected from university professors
The attendance sheet, attendance card and booklets are printed for every candidate
The answer sheets are collected and then scanned to the system.
The scores are only displayed when the scanning of the answer sheets has been completed and matched with the answer bank.
Faculty selection is the final process of the system, candidates selecting faculties based on their interested and every faculty has a unique number.
Then students' results are sent to universities on paper based and to the student management information system (SMIS) by soft copy.

As the process is very complex, in Fig. 1 we abstract the registration process of exam candidates. In this process, the MoHE needs to collect all the exam candidate profiles from schools via the MoE and register them in the computer-based system. Those who do not succeed in the current exam, need to reregister in the following year, and this might be done several times.

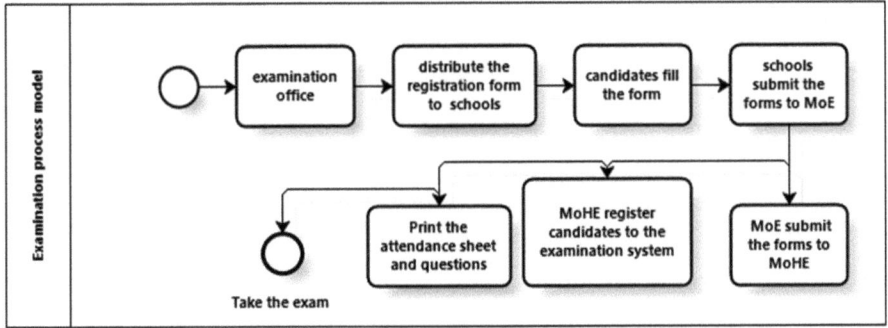

Fig. 1. Show the workflow of student's registration to examination system.

Figure 2 show the concept map of the relational model of the proposed e-examination system. There are relations between three modules: the registration system, the question bank and the post-examination system. It also demonstrates the relations between the components of each module. First of all, the MoE need to transfers the profile of the exam candidates via a web interface to the examination system, however currently they transfer the papers of profile of candidates to the MoHE. Then, based on candidates' attendance list, systems administrators print the booklets that have questions, the faculty number and the faculty selection list. When the exam is completed, all papers are

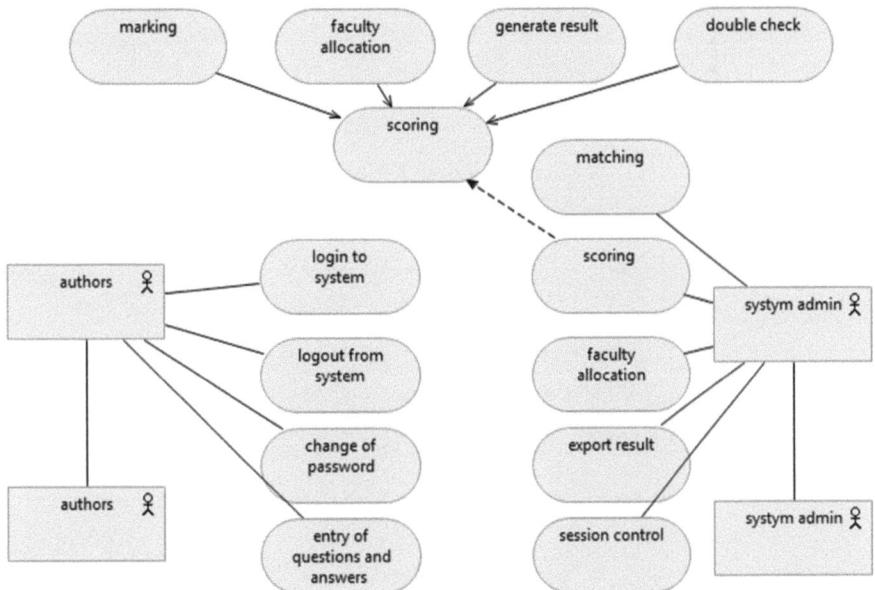

Fig. 2. Relation model of e-assessment system

collected by examiners and delivered back to the examination office in the MoHE. Then, after scanning the answer sheets with a machine of optical mark reader (OMR), after matching the answer sheet with the stored answers in the question bank, and after matching the faculty selection, the result of the exam will be available that can be exported in a PDF or an Excel sheet. At the same time, the examination system provides the option to active and take an online examination. The administrator can active the exam based on the attendance of candidates, and candidates can pass exam via web interface from the examination centers. They can see their score after submitting all answers, but the final result will be available only when all candidates has completed the exam.

3.4 Use Case Diagram

The university entrance exam is an assessment of school graduates who are candidates for higher education degrees. Figures 3 and 4 show the use case diagram of the university entrance exam, where three main organizations are involved. The MoE is responsible for schools who provide candidates for the exam. The MoHE organize and administrate the examination. The universities are the target for the candidates as well as the university professors provide questions and answers.

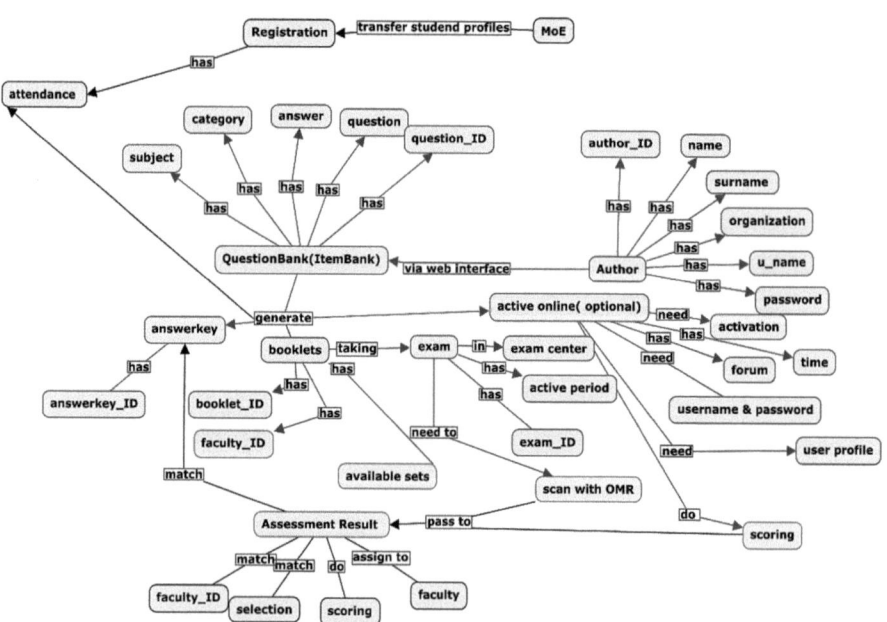

Fig. 3. Use case diagram of the question bank (Item Bank) and the assessment result

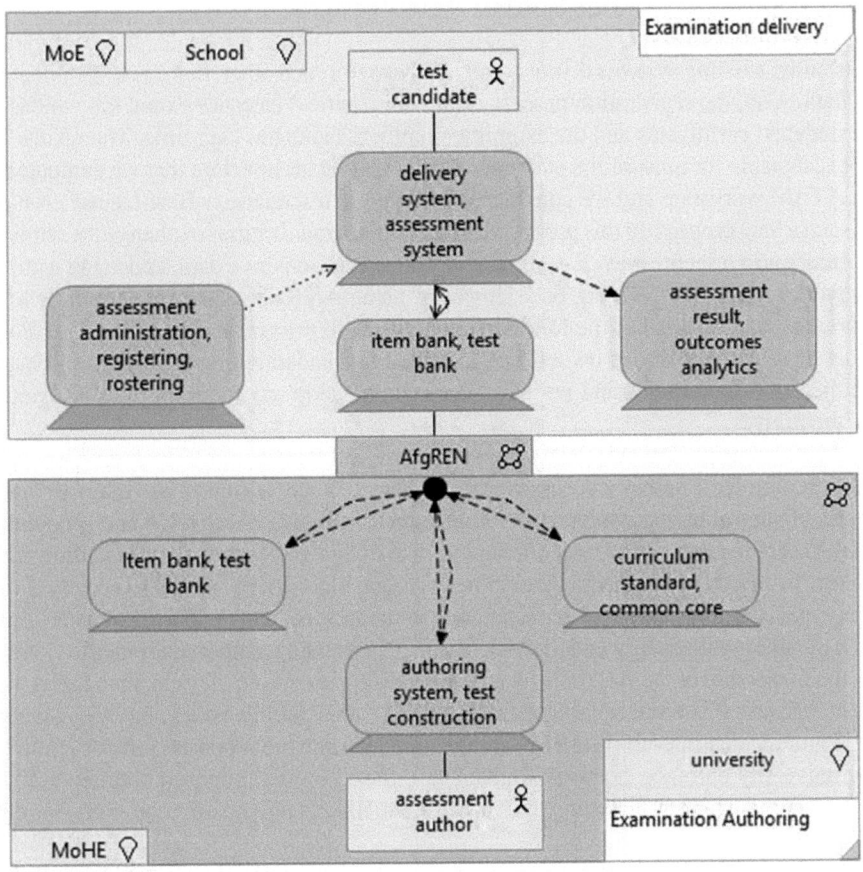

Fig. 4. Use case diagram of the university entrance exam based on AfgREN.

Figure 3 demonstrates the use case of a part of the system that is post-examination or result analysis section. There is more administration that is organized by the MoHE and authoring that is provided by university professors. Figure 4 adapted from QTI[4] illustrates the complete system. There are two main environments: examination authoring and examination delivery. Examination authoring section include authors and administrators, this part is operated by universities and by the MoHE. The examination delivery part is operated by AfgREN, the MoE and schools, this part is more for users rather than the administration.

---

[4] https://www.imsglobal.org/question/qtiv2p2/imsqti_v2p2_oview.html.

4 Proposed E-testing System Architecture

Currently, existing systems do not cooperate between each other, this causes duplication of data. Also, the registration process of the universities' entrance exam, the validation of students' certificates and the examination process take too long time. The results are not acceptable for most of the customer. Our proposed architecture that we evaluated in an ATAM workshop and we considered e-testing as a scenario, will solve the problem of such a long process. In this proposed architecture, organizations exchange the required documents in a secure way. E-testing always involves sensitive data, and using a public network for e-testing is a big risk. However, using AfgREN as the infrastructure of e-learning and e-testing has the following benefits, high privacy and security: AfgREN is a national network that is owned and managed by academic organizations, and based on their developed rules and policies. They can develop any required rules and policy by themselves.

More control: The AfgREN provides many different controls including the user access control, the network access control, the devices access control and many different levels of control by the implementation of authentication, authorization and accounting (AAA) server. Energy and cost efficiency: to maintain and secure all information technology in all educational organization needs much more energy and is too costly. Even it is especially difficult in Afghanistan, due to the lack of many experts in this area, the lack of sustainable energy and the lack of adequate funding. Improved reliability: based on the ownership of the AfgREN by an academic organization, they can manage as they want, because it is more reliable than a network that is public and owned by some else.

Figure 5 illustrates the AfgREN architecture [14] and the proposed infrastructure for the university entrance exam of Afghanistan. Based on the proposed architecture, the relevant organizations including the MoE, the MoHE and universities can exchange data

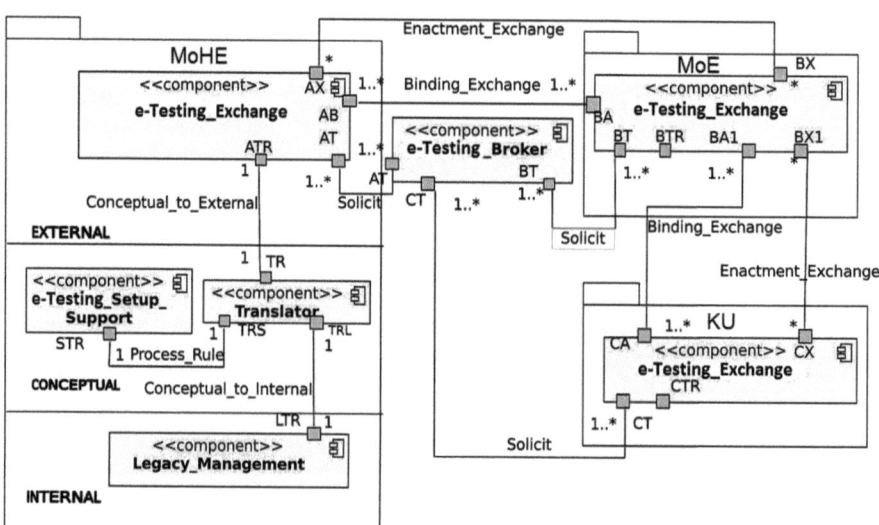

Fig. 5. NREN e-assessment architecture, adapted from [29]

in a secure, reliable, high performance network. The detail explanation of the architecture explained in [28]. Based on our proposed architecture, we use the current system, but a new version is needed, because the additional improvements need to be made.

The system should be upgraded to a new cloud-based adapted version
When the candidate log in with their own ID, the system should give a time that the administrator has assigned it.
When the time is near to finish, the system should inform the user.
We need to register a domain name for the system.
We need to introduce the cancour system to AfgREN and to AAA server.
Testing will be available only in the assigned time.

The proposed architecture can be implemented in any NREN, for example countries like Estonia that Estonian Education and Research Network (EENet[5]) already exists and student profiles already exist in schools system like OIS.

5 Conclusion and Future Work

In this research, we consider the university entrance exam which is the knowledge assessment of school graduates. This assessment is needed for the entrance to universities. However, our proposed architecture can be used for different types of assessments. In this research, we designed the university entrance exam based on AfgREN. This can be implemented also in other assessments, as well as in other countries. We applied design science research, and we use Requirement Bazaar to collect and analyze the requirements. We use the ATAM to evaluate the system.

We still need one more ATAM workshop specifically for the implementation of an e-assessment system with the NRENs architecture. We used the Requirement Bazaar format and concept and we conducted interviews with professors, with the administrative as well as technical staff who are involved in the university entrance exam in Afghanistan. However, we could not implemented the Requirement Bazaar in a proper manner, because they did not want to spend more time in the online system of the Requirement Bazaar. We explored the process model, relation model and use case of system and we adapted the e-testing system with AfgREN e-learning architecture that has interoperability, high security and high performance. The proper implementation of the Requirement Bazaar and another ATAM workshop for the e-assessment system in NREN will be our future work.

References

1. Gusev, M., Armenski, G.: E-assessment systems and online learning with adaptive testing. Stud. Comput. Intell. **528**, 145–183 (2014)
2. Ramos, N.: E-assessment of matlab assignments in moodle: application to an introductory programming course for engineers, pp. 1–9 (2010)

---

[5] https://www.eenet.ee/EENet/EENet_en.

3. Beetham, H., Sharpe, R.: Rethinking Pedagogy for a Digital Age. Routledge, Abingdon (2007)
4. Zhang, P., Wills, G., Gilbert, L.: IMS QTI engine on android to support mobile learning and assessment, pp. 1–8 (2010)
5. Hajjej, F., Hlaoui, Y.B., Ben Ayed, L.J.: Personalized and generic e-assessment process based on cloud computing. In: 2015 IEEE 39th Annual Computer Software and Applications Conference, pp. 387–392 (2015)
6. Singh, S., Chana, I.: A survey on resource scheduling in cloud computing: issues and challenges. J. Grid Comput. **68**, 1–48 (2016)
7. Dillon, T., Wu, C.W.C., Chang, E.: Cloud computing: issues and challenges. In: 24th IEEE International Conference on Advanced Information Networking and Applications (AINA), pp. 27–33 (2010)
8. Van Der Veldt, K., De Laat, C., Monga, I., Grosso, P., Dugan, J.: Carbon-aware path provisioning for NRENs. In: 2014 International Green Computing Conference (IGCC 2014) (2015)
9. Villalon, S.L.J., Hermosa, C.C.: The role of national research and education networks providing connectivity and advanced network services to virtual communities in collaborative R&E projects, p. 2630. The Mexican Case, CUDI (2016)
10. Muñoz-Calle, J., Fernández-Jiménez, F.J., Sierra, A.J., Martín-Rodríguez, Á., Ariza, T.: LTI for interoperating e-assessment tools with LMS. In: Caporuscio M., De la Prieta F., Di Mascio T., Gennari R., Gutiérrez Rodríguez J., Vittorini P. (eds.) Methodologies and Intelligent Systems for Technology Enhanced Learning. Advances in Intelligent Systems and Computing, vol 478. Springer, Cham (2016)
11. Miguel, J., Caballe, S., Xhafa, F., Prieto, J.: Security in online web learning assessment: providing an effective trustworthiness approach to support e-learning teams. World Wide Web **18**(6), 1655–1676 (2015)
12. Parshall, C.G., Harmes, J.C., Davey, T., Pashley, P.J.: Innovative items for computerized testing. In: van der Linden, W.J., Glas, C.A.W. (eds.) Elements of Adaptive Testing. Statistics for Social and Behavioral Sciences, pp. 215–230. Springer, New York (2010)
13. Arif, M., Illahi, M., Karim, A., Shamshirband, S., Alam, K.A., Farid, S., Iqbal, S., Buang, Z., Balas, V.E.: An architecture of agent-based multi-layer interactive e-learning and e-testing platform. Qual. Quant. **49**(6), 2435–2458 (2014)
14. Saay, S., Norta, A., Laanpere, M.: Towards an architecture for e-learning infrastructures on a national level: a case study of AfgREN. In: Gong, Z., Chiu, D., Zou, D. (eds.) Current Developments in Web Based Learning. LNCS, vol. 9584, pp. 98–107. Springer, Cham (2016). doi:10.1007/978-3-319-32865-2_11
15. Vaishnavi, V.K., Kuechler, B.: Design research in information systems. MIS Q. **28**(1), 75–105 (2010)
16. Nussbaumer, A., Kravcik, M., Renzel, D., Klamma, R., Berthold, M., Albert, D.: A framework for facilitating self-regulation in responsive open learning environments 1 introduction, no. 1, pp. 1–41 (2014)
17. Sottet, J.-S., Vagner, A., Frey, A.G.: Model transformation configuration and variability management for user interface design. In: Desfray, P., Filipe, J., Hammoudi, S., Pires, L.F. (eds.) MODELSWARD 2015. CCIS, vol. 580, pp. 390–404. Springer, Cham (2015). doi: 10.1007/978-3-319-27869-8_23
18. Kazman, R., Klein, M., Clements, P.: ATAM: Method for Architecture Evaluation. CMUSEI, vol. 4, p. 83 (2000)
19. Frezzo, D.C., Behrens, J.T., Mislevy, R.J.: Design patterns for Learning and assessment: facilitating the introduction of a complex simulation-based learning environment into a community of instructors. J. Sci. Educ. Technol. **19**(2), 105–114 (2010)

20. Clark, R.A., Pua, Y.H., Fortin, K., Ritchie, C., Webster, K.E., Denehy, L., Bryant, A.L.: Validity of the Microsoft Kinect for assessment of postural control. Gait Posture **36**(3), 372–377 (2012)
21. Beebe, M.: E-learning in Afghanistan, December 2010
22. Kumar, S., Gankotiya, A.K., Dutta, K.: A comparative study of moodle with other e-learning systems. In: 2011 3rd International Conference on Electronics Computer Technology (ICECT 2011), vol. 5, pp. 414–418 (2011)
23. Drechsler, A.: A postmodern perspective on socio-technical design science research in information systems. In: Donnellan, B., Helfert, M., Kenneally, J., VanderMeer, D., Rothenberger, M., Winter, R. (eds.) DESRIST 2015. LNCS, vol. 9073, pp. 152–167. Springer, Cham (2015). doi:10.1007/978-3-319-18714-3_10
24. Sorrentino, M., Virili, F.: Socio-technical perspectives on e-government initiatives. In: Traunmüller, R. (ed.) EGOV 2003. LNCS, vol. 2739, pp. 91–94. Springer, Heidelberg (2003). doi:10.1007/10929179_15
25. Schmidt, D.C.: Model-driven engineering. Comput. Comput. Soc. **39**, 25–31 (2006)
26. Rodrigues Da Silva, A.: Model-driven engineering: a survey supported by the unified conceptual model. Comput. Lang. Syst. Struct. **43**, 139–155 (2015)
27. Neto, R., Jorge Adeodato, P., Carolina Salgado, A.: A framework for data transformation in credit behavioral scoring applications based on model driven development. Expert Syst. Appl. **72**, 293–305 (2017)
28. Saay, S., Norta, A.: Towards an Architecture for e-Learning Infrastructures on a National Level: A Case Study of AfgREN, vol. 20. Elsevier (2016)
29. Norta, A., Grefen, P., Narendra, N.C.: A reference architecture for managing dynamic inter-organizational business processes. Data Knowl. Eng. **91**, 52–89 (2014)

A Review of Interactive Computer-Based Tasks in Large-Scale Studies: Can They Guide the Development of an Instrument to Assess Students' Digital Competence?

Leo A. Siiman[✉], Mario Mäeots, and Margus Pedaste

University of Tartu, Ülikooli 18, 50090 Tartu, Estonia
{leo.siiman,mario.maeots,margus.pedaste}@ut.ee

Abstract. We review interactive computer-based tasks from several large-scale ($n > 20,000$) educational assessments in order to better understand the advantages of these assessment items and how they can inform the development of computer-based items to assess students' digital competence. Digital competence is considered to be an essential competence in today's knowledge-based society and has been described in the DigComp framework as consisting of 21 individual competences grouped under 5 areas: information and data literacy, communication and collaboration, digital content-creation, safety and problem-solving. In the present paper we examine interactive computer-based tasks from three large-scale assessments and conclude by mapping constructs from these assessments to the DigComp framework. A look at this mapping provides an initial view of which aspects of the DigComp framework require the most attention in terms of developing interactive computer-based items for a potential instrument to assess students' digital competence.

Keywords: Digital competence · Computer-based assessment · Interactive computer-based tasks · Large-scale educational assessment

1 Introduction

New types of computer-based assessment items, such as interactive tasks, have started to appear in large-scale educational studies. Interactive computer-based tasks do not disclose all the information needed to solve a problem initially, but instead require students to explore a (simulated) environment and uncover additional information necessary to resolve a problem situation [1]. The PISA 2012 problem-solving assessment framework says that "Including interactive problem situations in the computer-based PISA 2012 problem-solving assessment allows a wide range of more authentic, real-life scenarios to be presented than would otherwise be possible using pen-and-paper tests. Problems where the student explores and controls a simulated environment are a distinctive feature of the assessment" [2]. The trend towards new assessment paradigms is enabled by the use of digital computers and may offer opportunities to better measure complex competences.

One very important competence – which in 2006 was acknowledged by the European Parliament and Council as one of eight key competences for lifelong learning necessary for personal fulfilment, active citizenship, social cohesion and employability – is digital competence [3]. Ferrari [4], in a report commissioned by the European Commission, proposed a DigComp framework consisting of 21 separate digital competences divided in 5 areas. The five areas according to a recent revision to the DigComp framework by Vuorikari, Punie, Carretero Gomez and Van den Brande [5] are: information and data literacy, communication and collaboration, digital content-creation, safety and problem-solving.

Although the DigComp framework offers a self-assessment grid with three proficiency levels, the items tend to be abstract and lack clarifying examples (e.g. *I can use a wide range of strategies when searching for information and browsing on the Internet*). Siiman et al. [6] found that terms thought to be synonymous in DigComp items were not readily understood to be similar by 6th and 9th grade students. In addition, a considerable concern with self-reports by young people is that it may actually provide misleading results. In the 2013 International Computer and Information Literacy Study (ICILS), 89% of eighth grade students answered a self-report questionnaire that they "feel confident to find information on the internet" yet on a standardized testing portion of the study only 23% achieved a proficiency level of sufficient knowledge and skills of ICT for information gathering and use [7]. Thus, in order to ensure a reliable measure of students' digital competence it is worthwhile to examine what we can learn from educational assessments that have gone through large-scale reliability and validity checks. Furthermore, since digital competence is naturally suited to be measured by computer assessment, our current focus on innovative interactive computer-based tasks is believed to be well-justified.

Three large-scale educational assessments are examined in this paper: the Programme for International Student Assessment (PISA), Technology & Engineering Literacy assessment (TEL), and the International Computer and Information Literacy Study (ICILS). Our aim is to examine the affordances of interactive computer-based tasks in each of these assessments and conclude the paper by mapping relevant constructs from the examined assessment tasks to competences in the DigComp framework. The completeness or incompleteness of this mapping will provide us with an overview of where the most work needs to be done with regards to developing an instrument to assess digital competence.

2 PISA

The Programme for International Student Assessment (PISA) is an international survey sponsored by the Organisation for Economic Co-operation and Development (OECD) that began in the year 2000 with the participation of 43 countries/economies and since then takes place every three years [8]. PISA aims to evaluate to what extent 15-year-old students can "apply their knowledge to real-life situations and be equipped for full participation in society". With a focus on testing broad concepts and skills that allow knowledge to be applied in unfamiliar settings, PISA provides a framework that can guide assessments into new competency domains. Although PISA has assessed multiple domains (mathematics, science, reading, individual problem-solving, collaborative problem solving, financial literacy), we choose to examine interactive computer-based tasks from the 2009 digital reading, the 2012 (individual) problem solving, and the 2015 collaborative problem solving assessments because these domains were judged to be the most relevant for developing an instrument to measure digital competence.

2.1 PISA 2009 Assessment of Digital Reading

In 2009, PISA offered an optional test, in addition to its main test, to assess the ability of students to navigate and evaluate digital information. About 36,500 15-year-olds from 19 countries/economies participated in this assessment [9]. While digital reading is in many ways similar to print reading, the PISA 2009 assessment framework distinguished electronic (i.e. digital) text as "synonymous with *hypertext*: a text or texts with navigation tools and features that make possible and indeed even require non-sequential reading. In the electronic medium, typically only a fraction of the available text can be seen at any one time, and often the extent of text available is unknown" [10, p. 27]. Moreover, the framework pointed out that "Gathering information on the Internet requires skimming and scanning through large amounts of material and immediately evaluating its credibility. Critical thinking, therefore, has become more important than ever in reading literacy" [10, p. 22]. Taking into consideration the non-linear navigation features of digital text and the need to critically evaluate digital texts, the final report on the results of the PISA 2009 assessment of digital reading defined the highest level of proficiency in digital reading as follows:

Tasks at this level typically require the reader to locate, analyse and critically evaluate information, related to an unfamiliar context, in the presence of ambiguity. They require generating criteria to evaluate the text. Tasks may require navigation across multiple sites without explicit direction, and detailed interrogation of texts in a variety of formats [9, p. 46].

Only 8% of students performed at the highest level of digital reading proficiency across the 16 participating OECD countries. Furthermore, analysis of process data generated by students (e.g. sequence of page navigation to complete a task) showed that proficient digital readers tended to minimize visits to irrelevant pages, efficiently located and compared critical information from different pages when necessary, and spent more time on complex tasks than on simpler tasks [9].

An example of an interactive digital reading task is shown in Fig. 1. The task consisted of three questions which taken together required students to perform five actions: (1) scroll through pages to find relevant information, (2) avoid distracting and irrelevant links and side menus, (3) integrate and interpret text, (4) reflect on and evaluate two pages in terms of credibility/trustworthiness of information, and (5) synthesis information from two pages. The interactive task displayed a simulated web browser and navigation functionality that was similar to many actual web browsers and online web pages.

Fig. 1. Example of a publicly released item, the *SMELL* unit, from the PISA 2009 Digital reading assessment. (Source: https://www.oecd.org/pisa/pisaproducts/pisa-test-questions.htm)

2.2 PISA 2012 Assessment of Problem Solving

In 2012, PISA offered an optional assessment for problem solving, of which 85,000 15-year-olds from 44 participating and economies took part in [11]. PISA 2012 framework defined (individual) problem-solving competence as:

Problem-solving competency is an individual's capacity to engage in cognitive processing to understand and resolve problem situations where a method of solution is not immediately obvious. It includes the willingness to engage with such situations in order to achieve one's potential as a constructive and reflective citizen. (OECD, 2013a).

The problem solving framework identifies four competences or processes in problem solving (*Exploring and understanding, Representing and formulating, Planning and executing, Monitoring and reflecting*) [11]. Results from the assessment of problem solving distinguish the two highest proficiency levels (5 and 6) on problem-solving by noting that:

When faced with a complex problem involving multiple constraints or unknowns, students whose highest level of proficiency is Level 5 try to solve them through targeted exploration, methodical execution of multi-step plans, and attentive monitoring of progress. In contrast, Level 6 problem-solvers are able to start by developing an overall strategic plan based on a complete mental model of the problem [11].

An example of an interactive problem solving task is shown in Fig. 2. The task consisted of four questions which taken together required students to explore a technological device to understand and explain how it functions, to bring the device to a target state by planning and executing a specific sequence of actions, to form a mental representation of the way the whole system works, and to reconceptualise the way the device works to propose improvements to the device. The interactive task displayed a simulated MP3 player device in which three buttons on the device could be pressed. A reset button was also available to allow the student to return the device to its original state. The current state of the device could be determined by colour changes of elements in the MP3 player's display window. Scoring of one of the four questions in this task relied on process data captured by the computer (i.e. the number of clicks used) and students were instructed to complete this question using as few clicks as possible.

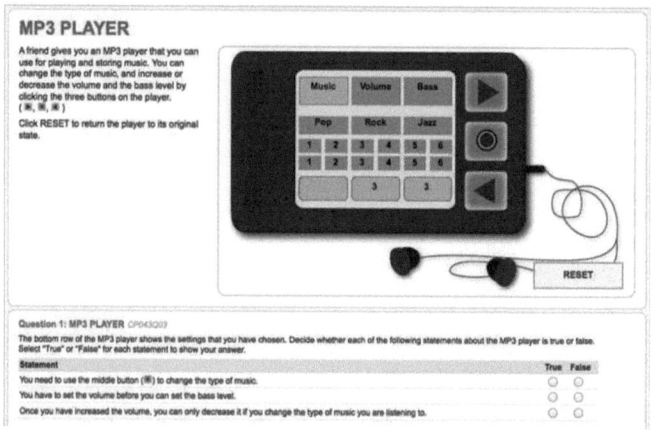

Fig. 2. Example of a publicly released item, the *MP3 PLAYER* unit, from the PISA 2012 Creative problem solving assessment. (Source: https://www.oecd.org/pisa/pisaproducts/pisa-test-questions.htm)

2.3 PISA 2015 Assessment of Collaborative Problem Solving

In 2015, PISA planned to administer a collaborative problem solving assessment. This competency is defined by in the draft collaborative problem solving framework as:

Collaborative problem solving competency is the capacity of an individual to effectively engage in a process whereby two or more agents attempt to solve a problem by sharing the understanding and effort required to come to a solution and pooling their knowledge, skills and efforts to reach that solution [12].

The above definition allows the framework to identify three collaborative competences (*establishing and maintaining a shared understanding, taking appropriate action to solve the problem, establishing and maintaining team organization*), which taken together with the four (individual) problem solving competences provide a comprehensive assessment of collaborative problem solving competency.

An example of a collaborative problem solving task is shown in Fig. 3. The task in general requires students to "chat" with simulated agents or team members to solve a given problem. Assessment items are often multiple-choice chat options presented to students from which they must select the single most appropriate choice. Along with chat interactions there is another area of the screen where stimulus material for the task is displayed. To ensure that incorrect/non-optimal choices do not hinder a students' ability to successfully complete subsequent items, the agents will assist in providing the required information to continue to progress through the task.

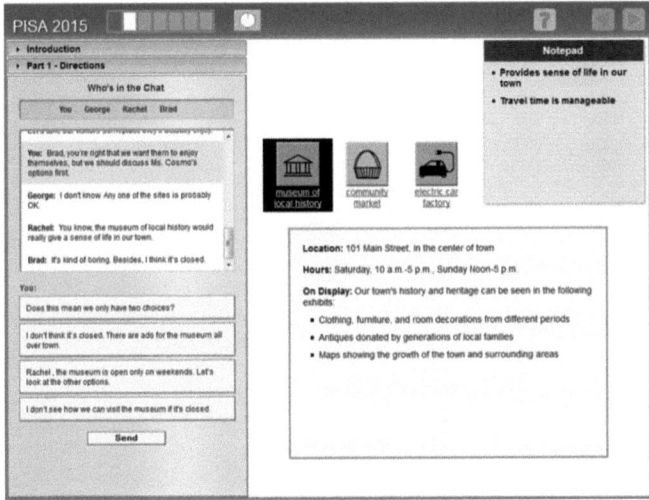

Fig. 3. Example of a publicly released item, the *VISIT* unit, from the PISA 2015 Collaborative problem solving assessment. (Source: https://www.oecd.org/pisa/pisaproducts/pisa-test-questions.htm)

3 TEL

In 2014, the National Assessment of Educational Progress (NAEP) administered the Technology and Engineering Literacy (TEL) assessment to 21,500 U.S. eighth grade students (http://www.nationsreportcard.gov/tel_2014/). According to the TEL framework, technology and engineering literacy is defined broadly as:

Technology and engineering literacy is the capacity to use, understand, and evaluate technology as well as to understand technological principles and strategies needed to develop solutions and achieve goals [13].

The framework identifies three major content areas, similar to competence areas, to be assessed (*Technology and Society, Design and Systems,* and *Information and Communication Technology*), along with three practices or ways of thinking when responding to TEL test items (*Understanding Technological Principles, Developing Solutions and Achieving Goals, Communicating and Collaborating*). The ICT content area in the framework is further divided into five subareas: *Construction and exchange of ideas and solutions, Information research, Investigation of problems, Acknowledgment of ideas and information,* and *Selection and use of digital tools.*

The TEL framework calls for the use of interactive computer-based scenarios that can, as far as possible, reflect tasks that might be done in society, and ask students to perform actions using simulated tools to solve problems in complex scenarios that reflect realistic situations. Tasks were designed so that students who answered incorrectly could still continue to subsequent questions and have an opportunity to answer them.

An example of an interactive TEL task is shown in Fig. 4. The task involves creating website content to communicate a convincing argument. Students had to identify relevant multimedia elements for presenting a message to a target audience, organize multimedia elements, and provide constructive and balanced feedback about a multimedia element.

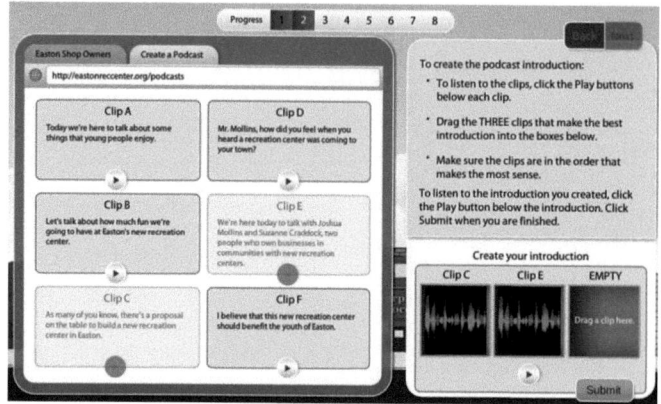

Fig. 4. Example of a publicly released item, the *Recreation Center* task, from the TEL 2014 assessment. (Source: http://www.nationsreportcard.gov/tel_2014/#tasks/reccenter)

4 ICILS

In 2013, the International Association for the Evaluation of Educational Achievement (IEA) conducted the International Computer and Information Literacy Study (ICILS) to assess the computer and information literacy skills of 60,000 eighth grade students ($M = 13.5$ years) from 21 education systems across the world [14].

Computer and information literacy is defined in the ICILS framework as "Computer and information literacy refers to an individual's ability to use computers to investigate, create, and communicate in order to participate effectively at home, at school, in the workplace, and in society" [14, p. 17]. The conceptual structure of the ICILS framework is comprised of two main constructs or strands (*Collecting and managing information, Producing and exchanging information*). The first construct contains three sub-constructs or aspects (*Knowing about and understanding computer use, Accessing and evaluating information, Managing information*) and the second construct consists of four sub-constructs or aspects (*Transforming information, Creating information, Sharing information, Using information safely and securely*).

An example of an interactive ICILS task is shown in Fig. 5. The task involves sharing information via email, navigating to a URL in a web browser, changing the share settings of a collaborate workspace website, creating user accounts on the collaborative workspace, evaluating the trustworthiness of an email message, and finally creating a poster using basic presentation software.

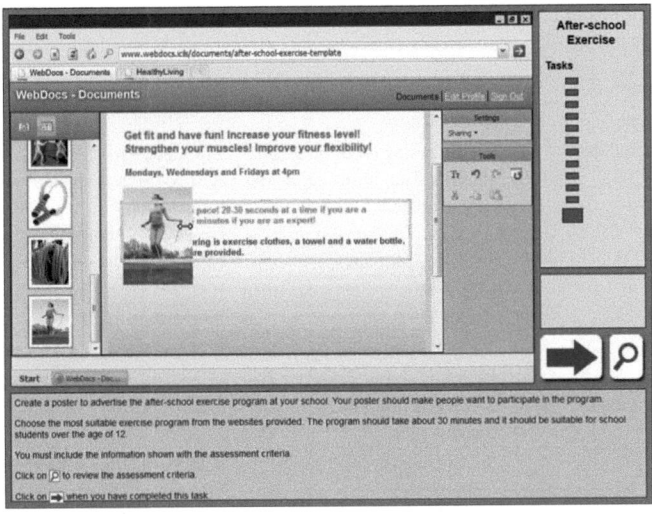

Fig. 5. Example of a publicly released item, the *Recreation Center* task, from the ICILS 2013 assessment. (Source: http://www.iea.nl/icils_2013_example_module.html)

5 Mapping of Constructs from the PISA, TEL and ICILS Large-Scale Assessments to the DigComp Framework

From the frameworks descriptions of the various constructs found in the PISA, TEL and ICILS assessments it was possible to map relevant constructs to the competence areas and competences of the DigComp framework. Most importantly, the definition and meaning of an assessment construct was examined to see if terminology and/or meaning overlapped with any of the 21 DigComp competences. Two of the authors of this study independently performed the mapping and demonstrated a *substantial* inter-rater agreement ($\kappa = 0.65$). Table 1 presents the results of the mapping.

It can be seen that not all of the DigComp competences could be represented by constructs found in the examined large-scale educational assessments in which interactive computer-based tasks are prominently represented. There are 10 of the 21 digital competences that were not mapped at all. A recent review of 38 instruments for assessing primary and secondary school students' ICT literacy also showed an uneven mapping against the DigComp framework [15]. Thus, it appears that some DigComp competences may require the development of new assessment items. However, it is reasonable to assume that interactive computer-based tasks can be developed to measure the underrepresented or missing DigComp competences. A closer look at the definition of specific DigComp competences can help foster thinking of potential interactive scenarios that provide evidence of the skill the given competence is presumed to elicit. An assessment instrument that fully measures all aspects of digital competence according to the DigComp framework would be a significant advance in providing citizens and young people with a reliable way to gauge their preparedness to deal with rapidly changing technology in today's knowledge-based society.

Table 1. Mapping of constructs from the PISA, TEL, and ICILS large-scale assessments to the DigComp framework.

DigComp framework		Constructs from the PISA, TEL, or ICILS assessments
Competence area	Competences	
1. Information and data literacy	1.1 Browsing, searching and filtering data, information and digital content	PISA 2009 Assessment of digital reading ICILS – Accessing and evaluating information TEL ICT –Information research
	1.2 Evaluating data, information and digital content	PISA 2009 Assessment of digital reading ICILS – Accessing and evaluating information
	1.3 Managing data, information and digital content	ICILS – Managing information
2. Communication and collaboration	2.1 Interacting through digital technologies	NO MATCH
	2.2 Sharing through digital technologies	ICILS – Sharing information TEL ICT –Construction and exchange of ideas and solutions
	2.3 Engaging in citizenship through digital technologies	NO MATCH
	2.4 Collaborating through digital technologies	PISA 2015 Assessment of collaborative problem solving
	2.5 Netiquette	NO MATCH
	2.6 Managing digital identity	NO MATCH
3. Digital content creation	3.1 Developing digital content	ICILS – Creating information
	3.2 Integrating and re-elaborating digital content	ICILS – Transforming information
	3.3 Copyright and licenses	TEL ICT –Acknowledgment of ideas and information
	3.4 Programming	NO MATCH
4. Safety	4.1 Protecting devices	NO MATCH
	4.2 Protecting personal data and privacy	ICILS – Using information safely and securely
	4.3 Protecting health and well-being	NO MATCH
	4.4 Protecting the environment	NO MATCH
5. Problem solving	5.1 Solving technical problems	NO MATCH
	5.2 Identifying needs and technological responses	ICILS – Knowing about and understanding computer use TEL ICT –Investigation of problems
	5.3 Creatively using digital technologies	PISA 2012 Assessment of problem solving
	5.4 Identifying digital competence gaps	NO MATCH

6 Conclusion

In this paper we examined interactive computer-based tasks from several large-scale educational assessments in order to better understand the potential advantages of using items from those studies to assess the digital competence of students'. Our results of mapping constructs from the large-scale assessment to competences in the DigComp framework reveal that several DigComp competences are not represented, and thus there may be a need to develop new interactive computer-based items to fully assess all aspects of students' digital competence according to the DigComp framework.

Acknowledgments. This study was funded by the Estonian Research Council through the institutional research funding project "Smart technologies and digital literacy in promoting a change of learning" (Grant Agreement No. IUT34-6).

References

1. Greiff, S., Holt, D.V., Funke, J.: Perspectives on problem solving in educational assessment: analytical, interactive, and collaborative problem solving. J. Probl. Solving 5(2), 71–91 (2013)
2. Organization for Economic Co-operation and Development. (OECD). PISA 2012 Assessment and Analytical Framework: Mathematics, Reading, Science, Problem Solving and Financial Literacy. OECD Publishing, Paris (2013a)
3. European Parliament and the Council. Recommendation of the European Parliament and of the Council of 18 December 2006 on key competences for lifelong learning. Official J. Eur. Union L394/310 (2006)
4. Ferrari, A.: DIGCOMP: A Framework for Developing and Understanding Digital Competence in Europe. Publications Office of the European Union, Luxembourg (2013)
5. Vuorikari, R., Punie, Y., Carretero Gomez, S., Van den Brande, G.:. DigComp 2.0: The Digital Competence Framework for Citizens. Update Phase 1: The Conceptual Reference Model. Luxembourg Publication Office of the European Union. EUR 27948 EN (2016). doi: 10.2791/11517
6. Siiman, L.A., et al.: An instrument for measuring students' perceived digital competence according to the DIGCOMP framework. In: Zaphiris, P., Ioannou, A. (eds.) LCT 2016. LNCS, vol. 9753, pp. 233–244. Springer, Heidelberg (2016). doi:10.1007/978-3-319-39483-1_22
7. Fraillon, J., Ainley, J., Schulz, W., Friedman, T., Gebhardt, E.: Preparing for Life in a Digital Age - The IEA International Computer and Information Literacy Study International Report. Springer, New York (2014)
8. Organization for Economic Co-operation and Development (OECD). Measuring Student Knowledge and Skills: A New Framework for Assessment. OECD Publishing, Paris (1999)
9. Organization for Economic Co-operation and Development (OECD). PISA 2009 Results: Students on Line: Digital Technologies and Performance, vol. VI. OECD Publishing, Paris (2011)
10. Organization for Economic Co-operation and Development (OECD). PISA 2009 Assessment Framework– Key Competencies in Reading, Mathematics and Science. OECD Publishing, Paris (2009)
11. Organization for Economic Co-operation and Development (OECD). PISA 2012 Results: Creative Problem Solving: Students' Skills in Tackling Real-Llife Problems, vol. V. OECD Publishing, Paris (2014)

12. Organization for Economic Co-operation and Development (OECD). PISA 2015 draft collaborative problem solving framework (2013b). https://www.oecd.org/pisa/pisaproducts/Draft%20PISA%202015%20Collaborative%20Problem%20Solving%20Framework%20.pdf
13. National Assessment Governing Board (NAGB). Technology and Engineering Literacy Framework for the 2014 National Assessment of Educational Progress. Author, Washington, DC (2013)
14. Fraillon, J., Schulz, W., Ainley, J.: International Computer and Information Literacy Study: Assessment Framework. IEA, Amsterdam (2013)
15. Siddiq, F., Hatlevik, O.E., Olsen, R.V., Throndsen, I., Scherer, R.: Taking a future perspective by learning from the past – a systematic review of assessment instruments that aim to measure primary and secondary school students' ICT literacy. Educ. Res. Rev. **19**, 58–84 (2016)

Design and Development of IMS QTI Compliant Lightweight Assessment Delivery System

Vladimir Tomberg[1(✉)], Pjotr Savitski[1], Pavel Djundik[2], and Vsevolods Berzinsh[2]

[1] School of Digital Technologies, Tallinn University, Tallinn, Estonia
{vladimir.tomberg,pjotr.savitski}@tlu.ee
[2] Estonian Entrepreneurship University of Applied Sciences, Tallinn, Estonia
thexpaw@gmail.com, sevaberz@gmail.com

Abstract. While testing students is a popular pedagogical evaluation approach in both formal and informal learning, there still is a lack of usable, interoperable and open source tools that could have enough motivating affordances for instructors. Another issue with the testing tools is a poor support of the question and test interoperability standards. Such standards exist for already seventeen years, however, the tools which support these standards are either big-scale proprietary products or are not easy to learn and use.

One of the reasons, why exchange of standardized tests and questions still is not a common place in the everyday pedagogical practice, especially outside of universities, is an absence of the high quality lightweight, open source tools.

The current paper sheds some light on the development issues of the interoperable assessment tools by using available open source libraries as the base components for the system core. We target pedagogical scenarios for testing in the informal learning, outside of universities. The study can be useful for designers and developers, who want to start development of standard-compliant software for e-assessment.

Keywords: Formative assessment · Assessment delivery system · Development · IMS QTI

1 Introduction

If one course instructor or trainer will look nowadays for a free and mobile-friendly assessment tool, which supports the latest standards for exchange of tests, and which could be easy to install, host, maintain, and use, and which can be promptly used for fast formative assessment, she will probably find some old software project, compatible with the old version of some development platform or operating system, and abandoned by developers a long time ago. Another alternative will be a product that is full of excessive functionalities, which is hard to install and understand and which is not always compliant with the most recent version of specifications. Most likely that product will not be free of charge nor open source. One more possibility would be to find a product that is being offered as a service with high chances of commercial interest.

IMS Question & Test Interoperability Specifications are available since 1999. During 17 years they had two major and several minor releases and nowadays the QTI is the only well-known and widely used standard for exchange of question and test data between heterogeneous assessment systems. The main purposes are described in the specification's body:

- To provide a well-documented content format for storing and exchanging items and tests, independent of the test construction tool used to create them;
- To support the deployment of item banks across a wide range of learning and assessment delivery systems;
- To support the deployment of items, item banks and tests from diverse sources in a single learning or assessment delivery system;
- To provide systems with the ability to report test results in a consistent manner [1].

There are many issues in the e-assessment domain that potentially can be solved by using standardization. Khaskheli [2] and Al-Smadi and Guetl [3] state that standards foster e-learning and e-assessment systems to ensure seven abilities: interoperability, reusability, manageability, accessibility, durability, scalability, and affordability.

While the purposes and advantages of QTI specifications are clear enough, it is not that simple to implement them well in the real life scenarios. There are several reports, which state that it is difficult to read the specifications, they are massive, have many design flaws and importing tests from unknown sources is complex [4–6].

Another issue is that the IMS QTI standard is poorly supported, both in LCMSs and authoring tools [7]. The list of QTI-supporting tools on Wikipedia[1] at the moment of writing of this paper consists of 40 applications from different vendors. However, almost half of them support the obsolete version 1.2 of QTI, which was released as far back as in 2002. Some applications support version 2.1 in a limited way, providing 'export only' or 'import only' functionalities. Many applications are abandoned by developers and are not developed further or supported for a long time. The similar situation can be found in another list of applications with IMS QTI support[2].

IMS Certified Product Directory[3] lists products that are guaranteed to meet the IMS standards for which they have passed testing. Currently, there are 22 products that belong to QTI section. Most of these products are proprietary, copyrighted, and often quite expensive products.

We have defined a list of shortages for the standard-compliant assessment tools, from the point of view of requirements demanded by assessment process in informal learning. These tools are:

- *Too complex.* They are intended for the universal functionality. While that can be considered useful for the formal learning, many of these functionalities may be obsolete for the informal learning context;
- *Resource consuming.* The wide spectrum of functionalities makes a software architecture more complex;

---

[1] https://en.wikipedia.org/wiki/QTI.
[2] http://www.liquisearch.com/qti/applications_with_ims_qti_support.
[3] https://www.imsglobal.org/cc/statuschart.cfm?show_all_nav_listCompliance3=1.

- *Hard to deploy and manage.* Large-scale, complex systems require the high level of competencies from the system administrators. Usually, course instructors do not have such competencies. Therefore, specially trained professionals are required for that;
- *Not easy to use.* The usability level of the most popular learning systems is not always the best and learning curve is rather steep;
- *Not mobile.* Most of the e-assessment systems have limited or no support for mobile devices;
- *Not open source.* There is no way to reuse libraries for development;
- *Not free.*

Obviously, IMS QTI is better supported by applications of the biggest players on the e-assessment market. Supporting the specifications in the small-scale development initiatives seems to be more challenging: there are many projects, which were started with a clear intention of supporting standards and then abandoned, without even getting close to becoming mature, feature-complete and having stable release version.

From one side, the specifications provide instructors with potential for a great level of flexibility, allowing creation, exchange, and reuse of tests and questions. From the other side, a notable level of continuous support of QTI can be expected mostly in the mainstream application products.

That situation is quite disappointing because standardization becomes useful only in case of using many separate applications from different vendors, by enabling interoperability. Major players on the market of learning management systems, like Blackboard and Moodle, already have every component needed for authoring, storing, and delivery to students. Some systems are self-sufficient and do not really need the compliance with the standard. Locking users into their ecosystem could be considered as a part of their business model. Contrariwise, small systems that provide the limited amount of separate functionalities, can survive only in case if they can exchange data with other systems. That could provide the instructors with flexibility, especially in informal, personalized, situated and mobile learning, where usage of a clumsy learning management system may hold learning process back.

That is also important from the point of view of freedom to design a learning process by involving different components from multiple different providers. It could allow the instructors to dynamically integrate any components into their teaching workflow [8]. At the same time, the small and modular applications are easier to include in the personal learning environments of the students.

While the described above situation exists on the market of e-assessment software for the last fifteen years, some small open source projects begin to appear on the market, providing components required for implementation of QTI-compliant software. We have examined several of them with a goal to develop a lightweight, open source assessment delivery system called *EzTest*, which could be promptly used for the flexible, mainly formative, assessment scenarios.

2 Requirements Design

For development of an IMS QTI compliant system, it is important to consider architecture and components of the assessment ecosystem defined in IMS QTI specifications, which provide opportunities for assessment workflow (Fig. 1).

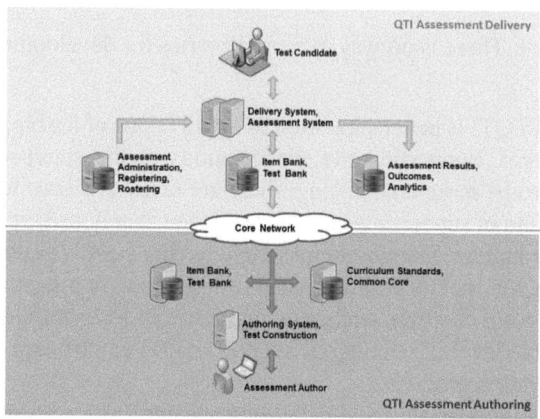

Fig. 1. Routes for assessment tests and items in QTI-enabled environment [1]

Specifications define two main parts of the assessment process: assessment authoring and assessment delivery. An *authoring system* and a *test bank* are the components of the QTI assessment authoring process. An *assessment delivery system* is the main component of the assessment delivery; it communicates with the test banks, provides an opportunity to manage and modify tests, manage students' groups, assign tests to candidates and provide access to learning analytics. Additional external services like Learning Management Systems (LMS) and learning analytics engines can communicate with the infrastructure by using QTI data exchange and necessary APIs.

In the current project, we have focused on design and development of the *assessment delivery system*. We consider that a lightweight assessment delivery system should have the ability to import QTI tests, prepared in the external applications, and provide a simple way for delivering the tests to candidates.

As a method for elaborating requirements, we have selected Scenario-Based Design [9–11]. During the design phase, we interviewed instructors, implemented several iterations of analyzing problems, designing user activities and interaction scenarios, rapid prototyping and evaluating.

2.1 Scenarios and User Activities

Below, we provide two scenarios based on the less formal learning context. They include two types of end-users: instructors and students.

Scenario 1. Instructor Mary and Summer School Students

Mary is an assistant professor at a faculty of psychology. She is invited as a mentor to one-week long university summer school. She has to provide a talk for the master students studying psychology, and she needs to have the better understanding of a level of preparation of the target audience. To investigate that, she wants to use a standardized test from a university LMS. Because the university LMS does not allow to register temporary summer school students, she uses EzTest application to import and combine a set of questions into a short survey. Then she collects emails from the students and assembles the new group of students within an instance of EzTest. After assigning the survey to the group, the test is activated and invites are sent to the students by email. Students answer the survey during the day. In the evening Mary sees the results and has an idea about the level of preparation of the group. Now Mary can prepare her talk by precisely targeting the current audience.

Scenario 2. Instructor Garry and Apprentice Student

Garry is a resident engineer in a telecommunication company, where he is responsible for a network infrastructure. Garry has an apprentice, who has to get three months of field experience in the company before getting a corresponding qualification. To improve skills of the student, Garry asks her to complete the professional self-tests within an instance of EzTest. For that, Garry imports the tests in QTI format from LMS of a company that provides network equipment into EzTest and assigns them to the apprentice. The student registers herself in EzTest and now she can continuously test herself by repeating assessment until she feels confident enough with the topic.

The following key activities were identified for the instructor in the scenarios above:

1. Authorization of the instructor in the system;
2. Import of questions from a file or package;
3. Creating a test from the imported questions;
4. Creating a user group by using email addresses of the users;
5. Assigning and activation of the test to the group;
6. Browsing results of the test's implementation.

In the scenarios above, the following activities were identified for the students:

1. Receiving email notification with a hyperlink to the assessment;
2. Optional registration in the system, if student wants to track her own progress;
3. Implementing the assessment;
4. Reviewing results;
5. Implementing the assessment again, if necessary.

The both sets of user activities are limited comparing to a traditional LMS, where these activities are usually accompanied by many others, which are necessary for supporting the formal learning process. By designing EzTest, we have followed a principle of Minimum Viable Product [12]. We provided only the minimal set of functionalities that allow implementation of the assessment process for our scenarios. Everything else that could be useful is omitted in the application to ensure simplicity and easiness to learn and use.

On the Fig. 2 three screenshots of the application prototype are shown. In particular, they show the mobile-oriented adaptive design approach which was implemented by using Bootstrap HTML, CSS, and JS framework (http://getbootstrap.com/).

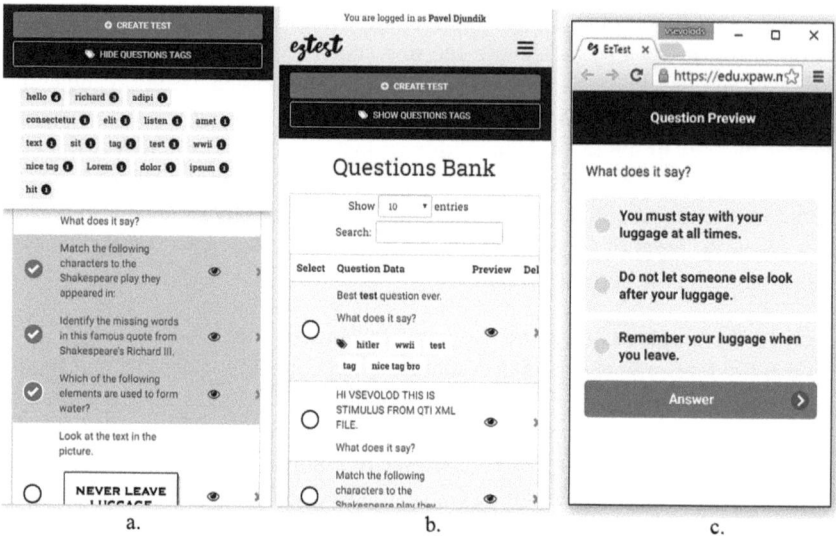

Fig. 2. Screenshots from EzTest application: a. instructors' question view, b. instructors' question bank view, c. student's assessment view

2.2 Defining Software Requirements

Based on the issues discussed in the previous section, we have defined the following list of functional requirements for the application. These requirements are used as a base for creating the Minimum Viable Product.

1. *QTI compatibility.* QTI specification version 2.2 defines 15 types of questions. Many of these types have complex visualizations and interactions and are rarely used. For the QTI version 1.2 there was a *'lite'* version of the specification, which supported only four types of questions, namely yes/no, true/false, likert scale, and multiple choice. In the later versions, development of QTI Lite was stopped, so for versions 2 and above the developers have to decide, which types of questions they would like to support. In our case, we have decided to support seven following types of questions: *matchInteraction*, *gapMatchInteraction*, *inlineChoiceInteraction*, *TextEntryInteraction*, *ExtendedTextInteraction*, *ChoiceInteraction*, and *OrderInteraction*. This set of questions' types enables the creation of typical questionnaires for surveys (yes/no, true/false, Likert scale questions, free text), as well as the creation of the simple tests for the assessments.
2. *Compatibility of the application with mobile devices.* Mobile Internet is the strong trend nowadays. That is especially important for the informal learning, which is "fundamentally mobile" [13]. In the first turn, we focus on an ability of the students

to implement online tests by using their mobile devices. Secondly, we challenged ourselves to achieve the smooth workflow for the teachers with mobile devices too.
3. *Minimum or no registration necessity for students.* In informal learning and vocational education, the tests often are conducted on demand and for one time only. In such cases, there is no necessity to register user accounts, which will never be used again. An absence of the user registration will result in better user experience because end users usually do not like to put additional effort into going through the registration process. That is especially true when the users are aware of one-time nature of the activity. In case if the user still wants to register, for example, for tracking results, the registration process has to be as simple as possible with the minimum of required effort from the users and a minimal amount of data requested.
4. *A possibility to easily import whole sets with tests or separate questions in QTI format and manage own collections on the level of separate questions.*
5. *Simple creation and management of lists of users.* We suppose that activation of the tests can be implemented by using an emailed link. In that case, only the list of emails is required, removing the need for students to register themselves. This makes the process of assigning a test to students much quicker.
6. *Simple and straightforward creation and assigning of the tests.* Selection of required questions from the local collection and their sequencing.
7. *Monitoring results for the specific assignment.*

By having in mind the functional requirements described above, we have also defined a short list of non-functional requirements that can be important to consider during design and implementation time.

1. *Supporting the last version of PHP, the PHP 7 series;*
2. *Program libraries being used should be well documented and still supported by their developers;*
3. *Using open source components;*
4. *Using open licenses.*

PHP was chosen because it is widely used, battle tested, has a wide user base and a lot of modules to use as building blocks for developed application. Wide adoption and big user base of the platform would ensure an ability to find developers and maintainers in the future.

Using the latest version ensures that the product is based on a platform that would still be supported for a meaningful period of time. Currently, PHP version 5.6 is the final release of the 5.x branch and version 7 is the future of PHP.

One more important aspect would be the ability to easily find affordable hosting solutions that would require little knowledge and experience for getting the service up and running. This allows keeping the upkeep costs to a minimum and running on virtually any hosting solution.

After identifying the main set of the functional and non-functional requirements and before starting development, we had to select the right technologies for the implementation of the application.

3 QTI Libraries

The problem with the development of QTI-compliant applications is a lack of the stable libraries for processing and visualization of tests and questions in QTI XML format. Most of the initiatives, like R2Q2[4] that was popular in 2006–2010th, were created as part of the universities' projects or by independent QTI enthusiasts. However, many projects have been abandoned by their developers and became unmaintained.

Considering the functional requirement 1 (QTI compatibility) and the non-functional requirement 2 (Program libraries being used should be well documented and still supported by their developers), we have examined three the most promising QTI processing libraries written in PHP available on GitHub today.

3.1 PHP-QTI

At first, we have examined PHP-QTI library of Michael Aherne from the University of Strathclyde[5]. Author notices, that the library software is under active development and some parts of the code can be not functional yet. We have examined the library and were not satisfied with the result.

We have met the first problem when tried to get the library with Composer. Even though the repository had the descriptor file called *composer.json* present, the library was inaccessible from the Composer itself.

We have also found out that the library can load only the full packets of QTI files. That does not conform to the requirement 4: *"A possibility to easily import whole sets with tests…"*.

Another problem was the lack of documentation and recent activity; it seemed that the initiative was abandoned before getting an attention from possible adopters and the community.

After identifying these limitations we have excluded PHP-QTI from the examination list.

3.2 QTI Software Development Kit for PHP

At the moment of this study, QTI SDK from the Open Assessment Technologies (OAT) had the highest and constant activities with the code on GitHub[6].

This library has some specific peculiarities. For example, it converts QTI XML files to HTML format with so-called "Goldilocks rendering". However, such approach doesn't use available interactive elements of control in web forms. Instead of use *<input type = "radio" checked>* a usual *ul/li* list with additional data attributes is used. To build something working on such format, one needs to use other libraries from OAT or to develop the own ones.

---

[4] http://www.r2q2.ecs.soton.ac.uk/overview/index.htm.
[5] https://github.com/micaherne/php-qti.
[6] https://github.com/oat-sa/qti-sdk.

Development of the own libraries doesn't seem to be a good idea because the code of the Development Kit is not well documented (non-functional requirement 2) and will demand a lot of resources to learn about its' structure.

After identifying circumstances mentioned above we have excluded the QTI Software Development Kit for PHP from the further considerations.

3.3 Learnosity QTI

The last library we have examined was Learnosity QTI[7]. For processing QTI XML files Learnosity QTI Converter software was selected. Developers of QTI Converter claim that it can read QTI XML files and convert them to JSON format, which is used by Learnosity suite. JSON format is easy to use for exchange in client-server conversations, especially when used with AJAX web development techniques for development of asynchronous web applications.

Storing data in JSON format allows adding support for new types of questions without changes in the database because all fields remain unchanged. That also allows working with JSON object at the server side for converting JSON data into objects by using native PHP functions *json_encode()* and *json_decode()*.

Using Learnosity QTI Converter should allow to transparently validate QTI XML files against errors and missing fields and to normalize data by simplifying the programming process.

Learnosity QTI Converter was initially built for the Learnosity suite. Therefore, names of types of questions in the Converter are different in comparison to IMS QTI. For example, the QTI *ChoiceInteraction* type is called in Learnosity *Multiple Choice Question*. These question types have the same properties and functionality. However, the names and data structures are different. The full Learnosity data structure is well documented and is easy to use for developers.

Considering all arguments stated above, as well as the current development of the Learnosity QTI Converter, it was selected as QTI processor for EzTest.

4 Found Issues

In the requirements, we aimed to build a solution that would be kept as small and simple as possible (Minimum Viable Product), using as many existing components as possible to achieve results at a fast pace.

Easy and convenient import of existing single QTI test and whole packages is at the heart of the application, as these are the building blocks used to create tests, being the central functionality of the application. Having a module that would cover as many needs as possible is crucial to the whole development process. Unfortunately, no module seemed to have the ability to provide the basic machinery to get the job done. While Learnosity QTI module has a relatively good documentation to get the development started and seems to be able to convert single items from the QTI XML representation

---

[7] https://github.com/Learnosity/learnosity-qti

rather well, it still has a quite a few features missing [14]. Some of those could be seen as crucial to the process, especially with bearing in mind the easy and convenient import requirement, like:

- Import and parsing of packages with questions in QTI format;
- Ability to deal with assets (e.g. images) included in question bodies.

With one of the most important base libraries not providing some of the crucial features, there is a need to implement these features as part of the system itself. For the specific Learnosity case the following additional implementations are required:

- *Ability to upload full packages.* As these packages are just ZIP archives, it is simple to implement using tooling provided by the base platform.
- *Parsing Manifest file of the package.* A manifest file has to be parsed in order to determine contained questions and select the suitable ones for additional processing by the Learnosity QTI.
- *Dealing with assets.* Some of the question bodies could include links to files, contained within the package. These would have to be processed and stored. Later on, there would be a need to replace the internal URLs with endpoints that serve these files. That would also require additional permission checks to limit the audience who have access to the assets.
- *Notifying a user about results of the import.* In a case of the multiple questions, the user will have to be notified about import process results, how many questions imported and how many were unsuitable. Also, there is a need to detect any parsing issues reported by the library and act on that.

There is another issue with QTI compatibility that could become important in the future. At the moment of writing, Learnosity QTI package only supports QTI version 2.1 and not the latest released version of 2.2 (2.2.1). The only possible way to fix that is to introduce that ability to the library itself, mostly relying on the will of current maintainers.

A possibility of assets being part of the question body (the simplest example would be an inclusion of images) could possibly affect mobile-friendliness requirement of the application. As handheld devices have limited amount of space and lower screen resolution, these files might have to be tailored in order not to substantially affect the user experience. An example would be to limit the size of images shown, possibly even based on the capabilities of the device at hand. Another issue would be the size of these assets with respect to limitations in mobile data services, as slower speeds and limitations of mobile data usage. A possible solution for the case of images could be either to generate a few sizes and transmit the one that is closest to capabilities of the mobile client or produce a version of an asset with required size at runtime.

At the moment of writing, EzTest had a status of alpha release. By continuing the development, the developers consider the issues mentioned above. The next planned phase is the testing of application with the end users. The code has the open license and the team is open for collaboration with other interested parties.

5 Conclusion

While the creation of IMS QTI Compliant Lightweight Assessment Delivery System is possible, it becomes a non-trivial task due to a few important factors. One of those, and probably the most important one, is the lack of QTI libraries that would be packed with all the required features, have thorough documentation and large community of adopters. This also requires digging deep into the QTI specifications to implement the missing features or create wrappers that would deal with the task and then use the functionality provided by the library to deal with the specific parts of the process.

It is tough to keep it simple and stick to the base of Minimum Viable Product without being distracted with excessive, though nice to have features.

Most of the libraries worth noting have their own peculiarities and specifics, often forcing one to be locked out in custom data format logic. Some of them are being developed by companies, which are using those as building blocks for their commercial products and solutions. This somewhat limits the adoption by the community and leaves the company in charge of further development, without much overview of the development timeline and features to be included over time. One of the possible indicators of that could be a lack of issues (GitHub specific indicator), hinting to the fact that the library is not widely adopted and would probably become unmaintained if the company ever stopped the development.

We suppose that in a case of availability of the carefully designed, well-documented and supported by developers open software library for processing and rendering of QTI files, it is possible to expect the appearance of a wide spectrum of different applications from the third-party developers. In turn, that could influence popularization of standardized test resources and repositories for their storage.

In the current paper, we have made an overview of issues related to the QTI-compliant assessment delivery systems only. However, the same libraries for processing and visualization of the QTI questions could also be reused in QTI authoring tools. In that case, the additional level of interoperability will be ensured.

References

1. IMS Global Learning Consortium: IMS Question & Test Interoperability Specification. (215) AD
2. Khaskheli, A.: Intelligent agents and e-learning (2004)
3. AL-Smadi, M., Guetl, C.: Service-oriented flexible and interoperable assessment: towards a standardised e-assessment system. Int. J. Contin. Eng. Educ. Life Long Learn. **21**(4), 289–307 (2011)
4. Piotrowski, M.: QTI—a failed e-learning standard? In: Handbook of Research on E-Learning Standards and Interoperability, pp. 59–82 (2011)
5. Gorissen, P.: Quickscan QTI: usability study of QTI for De Digitale Universiteit., Utrecht (2003)
6. Gorissen, P.: Quickscan QTI – 2006: Usability study of QTI for De Digitale Universiteit., Utrecht (2006)
7. Gonzalez-Barbone, V., Llamas-Nistal, M.: eAssessment: trends in content reuse and standardization. In: Proceedings of Frontiers in Education Conference (FIE), pp. 11–16 (2007)

8. Tomberg, V., Laanpere, M., Ley, T., Normak, P.: Enhancing teacher control in a blog-based personal learning environment. IRRODL **14**, 109–133 (2013)
9. Carroll, J.: Making Use: Scenario-Based Design of Human-Computer Interactions. MIT Press, Cambridge (2000)
10. Rosson, M.B., Carroll, J.M.: Scenario-based design, pp. 1032–1050 (2002)
11. Thew, S., Sutcliffe, A., Procter, R., de Bruijn, O., McNaught, J., Venters, C.C., Buchan, I.: Requirements engineering for e-science: experiences in epidemiology. IEEE Softw. **26**, 80–87 (2009)
12. Moogk, D.R.: Minimum viable product and the importance of experimentation in technology startups. Technol. Innov. Manag. Rev. **2**, 23 (2012)
13. Sharples, M., Taylor, J., Vavoula, G.: Towards a theory of mobile learning. In: Proceedings of mLearning, vol. 1, pp. 1–9 (2005)
14. Learnosity: QTI Conversion [DRAFT] - Learnosity Documentation. https://docs.learnosity.com/authoring/qti/index

Exploring a Solution for Secured High Stake Tests on Students' Personal Devices

Ludo W. Van Meeuwen(✉), Floris P. Verhagen, and Perry J. Den Brok

Eindhoven University of Technology, De Rondom 70, 5612 AP Eindhoven, Netherlands
L.W.v.Meeuwen@TUe.nl

Abstract. Digital testing requires software and hardware. A hardware option is to establish dedicated test halls. As an alternative, this paper describes a setup and first experiences of a pilot run on a proof of concept, to answer the questions: which conditions are needed to create a secure solution for high stake assessments via students' personal devices (e.g., notebook from students)?

To answer this question, a proof of concept was run based on an outline of a Secure Test Environment Protocol (STEP). This STEP was based on three shells of security: Prevention, Detection and Reaction. Prevention concerns the technical solutions to prevent the use of unauthorized sources during tests. Detection concerns the risks (i.e., chance and impact) and the possibilities to detect unauthorized behavior and security issues. Reaction gives insight in requirements needed for an organization when students' personal notebook are being used in taking high stake tests.

The preliminary conclusion from the proof of concept is that it is possible to use students' personal notebook in secured high stake tests if a STEP - comprising three shells of security - has been implemented. Suggestions for further development are given.

Keywords: Security in digital tests · Bring your own device · Controlling students' personal notebooks · Secure test environment protocol

1 Introduction

The access to specific software during testing is becoming a necessity to pursue an optimal validity of a test in the curriculum of a university, such as Eindhoven University of Technology (TU/e) as the focus of the present contribution. If for example the learning objectives of a course comprise the use of software (e.g., Matlab, SPSS), the software should be made available for students when they take the test of the course; when a large amount of writing is required to answer questions, text editing software can be more efficient (e.g., typing is faster than writing, digital editing is easier, grading is easier), etcetera. In line with findings of SURF [1, 2], Teacher experiences at the university show that students should be provided a secured digital test environment to avoid access to communication software (e.g., Skype, desktop sharing, drop-box, e-mail) and to avoid unwanted access to data on the hard disk and/or on the internet. Monitoring only seems insufficient to avoid fraud in digital assessment. The literature suggests four conditions to optimally prevent students to commit fraud: 1. the boundaries of what is acceptable should be communicated to the student, 2. situations conducive to fraud should be avoided (e.g., secured environment in digital tests), 3. Test situations should be monitored so cheating during testing can be detected, and 4. In the event of fraud, sanctions should be imposed [3, 4, cf. 5–7].

In order to create a secured digital testing environment, several educational institutes (e.g., in the Netherlands) established test halls equipped with computers that are fully run by the institute (e.g., University of Utrecht, Delft University). Next to the advantages (e.g., more experiences in the field, easier technology), the establishment of such halls are known to have disadvantages: The rooms are hardly or not available for other purposes than taking tests (hence, during testing period in use only), the investment and operational costs are high, and scaling is difficult in case of fluctuating (e.g., increasing) numbers of students.

Therefore, the desire of educational institutes is to use the students' personal device (i.e., notebook) more than in the current situation is the case. However, little knowledge is available about using students' personal notebooks during high stake tests [8, 9].

This paper explores a solution for secured high stake tests using students' personal devices to answer the questions: *which conditions are needed to create a secure solution for high stake assessments on students' personal device (e.g., notebook from students)?* Therefore this paper first introduces the three steps of a Secure Test Environment Protocol, based on an earlier performed desk research [8]: 1. Prevention, 2. Detection, 3. Reaction. Next it describes the experiences of a pilot run of a proof of concept [10]. Finally, the conclusions are drawn, based on the evaluation of that pilot.

2 Three Security Shells

Obtaining one hundred percent safety during high-stakes testing is an illusion. Therefore, the aim is to obtain an optimal Secure Test Environment Protocol (STEP). Such a protocol comprises three steps or shells (Fig. 1) [8, 11, 12]:

1. Prevention: Options for committing fraud should be averted as much as possible by the institution,
2. Detection: It should (pro-actively) monitor - known and potential - security leaks and last but not least,
3. Reaction: the institution should react when a security leak has been found or has been suspected.

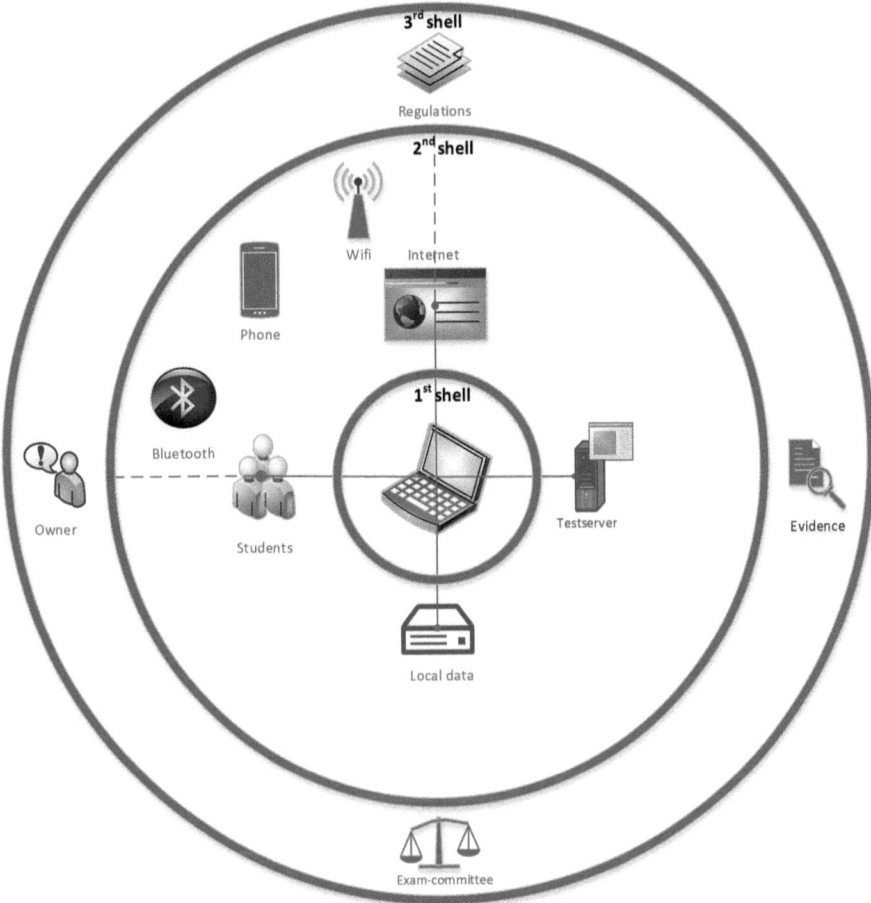

Fig. 1. Three security shells for a complete STEP: 1. prevention, 2. detection, 3. reaction.

3 Creating a Controlled Test Environment on Students' Notebooks in Shell One

The first step is to create shell one; how can students' personal notebooks be configured in such a way that basic security for STEP can be obtained?

3.1 Controlling Students' Notebooks

The use of personal notebooks implies that students can 'prepare' their notebook in order to by-pass security and/or monitoring software during testing. The prevention of such fraud during tests, starts with controlling the personal notebooks during tests. This control can be called the inner shell of the secured digital test environment.

An estimation of risks is needed to be able to define the best optimal solution for the notebook configuration during tests. Risk is defined as chances for fraud by the impact of the fraud. Impact can be *very high* (e.g., taking tests from outside the test hall), *high* (e.g., full communication possible), *medium* (e.g., extra information available via internet), or *relatively low* (e.g., access to local hard drive/limited information). The estimation of chance is depending on the a. skills from a cheater needed to modify the computer for cheating, b. time needed to cheat during the test, and c. time needed to invest prior to the test [8]. Verhagen and De Hesselle [13], made an estimation of the risks for fraud for mainly two scenarios (i.e., standard notebook with WiFi available and without WiFi).

There are almost endless possibilities when using software to by-pass security control, but there are possibilities to drastically reduce these options too. Hence, the design of the secured environment on personal notebooks (i.e., shell one) aims at diminishing these options. Taking control as early as possible in the booting process has the highest success rate (i.e., the later in the process, the more options and less complicated knowledge is required to by-pass security) [8]. Figure 2 outlines the booting process of a computer in four steps: First, the Video card, system board, and RAM memory are being addressed (i.e., BIOS start). High level computer knowledge is needed to change this part of the booting process. Verhagen [8] estimates that even students on a university of technology will be challenged to by-pass this process during testing, rare exceptions excluded. Second, the computer storage will be accessed to start the operating system (step 3; e.g., Windows) that allows applications (step 4; e.g., a browser) to run.

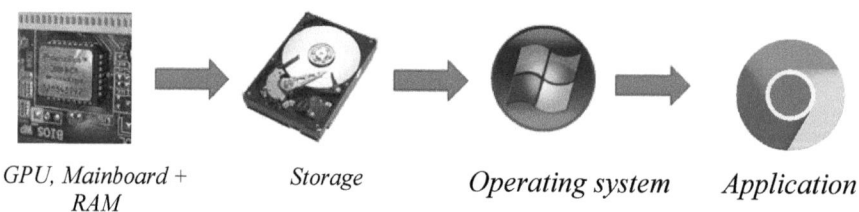

GPU, Mainboard + RAM   Storage   Operating system   Application

Fig. 2. Boot sequence.

Changes in storage, operating system or applications are easier to make and easier disguised. For example, a lock-down browser is an application that allows students to only have access to the test server. Because 'preparation' of a personal notebook is possible, manipulation by use of scripting within the storage or operating system could make it possible to have access to other applications while using the lock-down browser in such a way that detection by monitoring software is unlikely. In the event of additional network monitoring to such security solution, students could avoid detection of their fraud (e.g., by using other Wi-Fi connections like smartphones).Therefore, control over the personal notebooks should be taken over right before the moment the notebooks' storage is activated. The use of a virtual machine makes it possible to virtually imitate a physical machine. In theory, this could make it possible that the boot, controlled by the institution, is running on an operating system that is controlled by the student with all potential possibilities of fraud that can be imagined. It is hard for applications to detect the use of a virtual machine. However, Linux® has possibilities of detecting all the common virtualization programs.

3.2 Technically Three Options

The study by Verhagen [8] was focused at the situation of the TU/e where approximately 85% of the Bachelor Students are using the same sets of notebooks, selected each year by the institution. This allowed the focus for the pilot to work on the small variety of notebooks that the majority of the students have.

The institutionally controlled test environment on the personal notebooks should continue the boot process from the moment that the notebooks' storage would be addressed (see Fig. 2). Three options were defined for high stake tests to get an institutional controlled operating system (and applications) running on personal notebooks during tests.

1. Option A: Bootimage via UTP (wired) network, hence wired connection with boot- and test server.
2. Option B: Boot image via USB-stick, connection with test server (wired or Wi-Fi) (Fig. 2).
3. Option C: Boot image via Wi-Fi, hence, connection with boot- and test server via Wi-Fi.

Solutions A, B and C are similar as they all use the idea of a boot image that takes control over the personal notebooks. The difference is the way of distributing the image to the notebooks: A. a wired connection, B. a USB-stick, and C. via Wi-Fi.

Next to their risk analysis, Verhagen and De Hesselle [13] made an estimation of the potential control of the situation when the options A, B, and C (e.g., including the USB-stick solution) compared to the use of a lock-down browser.

In short, options A, B and C are almost equally secure. Option C. was excluded: The Wi-Fi capacity will not be sufficient in the coming few years to allow large groups of (say approx. 300) students to simultaneously download their secured image (including the download of software as Matlab or SPSS). Option A. was excluded too: A wired connection with a boot server means preparing a test-hall with wired internet connections and the installation and connection of several boot servers. The aim of this study was to avoid such investments. Option B, with the boot image via a USB-sticks and Wi-Fi connection to the test server turned out to be the best solution (Fig. 3). A proof of concept has been set up and described by Goossens and Koedam [10] using a Linux image on a 2-Euro USB-stick.

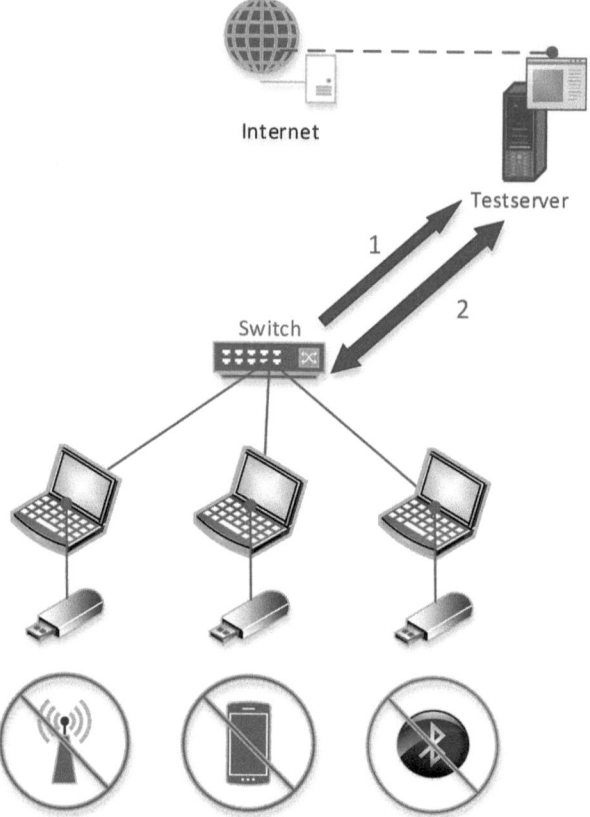

Fig. 3. Outline controlled test environment on personal notebooks.

4 Detection and Reaction

The USB-solution prevents students to use their notebooks for (non intended) communication during tests. As mentioned above, next to prevention, also detection and reaction are crucial aspects of a secured environment [11, 12]. In the desk research [8], prevention, detection and reaction are shown as three shells of the secured environment (Fig. 1). The inner shell represents the technical solution like solutions A, B, and C, mentioned above. The second shell represents the detection and the third shell represents the reaction.

4.1 Detection; Shell 2

The proctor will be an important factor in the security of tests. The security of the USB-solution is only useful when test taking takes place in a secured test hall. Monitoring increases chances of being caught when committing fraud and can help in detecting security leaks. This implies that proctors should be trained to be able to detect fraud in a digitalized situation. Furthermore, they should check for (wireless) connections (e.g., smart phone, smart watch, dongle) like they do with paper based exams. However, these checks are not at all security proof since the existence of a broad spectrum of wireless communication protocols.

A continuous way of monitoring using (HD) camera's is expected to work in a preventive way, and can make it easier later on by the examination committee to determine whether suspected fraud has occurred or not.

Logging the network traffic can be useful. However, experience shows that analyzing the logs can be a labor intensive job. So large scale network logging particularly is useful when the logs can be analyzed automatically. Preferable, notifications of suspicious network-use are generated automatically.

4.2 Reaction; Shell 3

"For the detection process to have any value there must be a timely response" ([11], p. 5.). This implies that ownership and processes must be arranged in case of security attacks [12]. These attacks could come from both inside and outside the organization. However, for now we focus at the danger from inside the organization (i.e., attempts of fraud by students).

The standard procedures in cases of fraud in paper based tests should be applied for digital testing too. In case of detection this means: writing an official report, safeguard the evidence (e.g., time stamps for data logs and video recordings, screen casts, etc.), communicating the case to the institutional body that is responsible for deciding on sanctions (e.g., examination committee).

In case the security of the first shell has been penetrated, an investigation should always be done in order to inspect the damage of the shell and plan reparation and improvement of the shell. This is not the task of an examination committee. This task should be assigned to the "owner" of the digital environment (e.g., software, hardware, regulations, and procedures).The impact of a single student that bypasses security is high, but easier to control than a scenario where a more general by-pass of the security is found that can work for a large group/all students.

Independent of fraud prevention, it is important to reduce the possibility for students to rationalize fraudulent behavior. A cat-and-mouse-game will probably start if prevention of fraud would be rule based only. An approach could be to express trust to students that they will become a scientist that will follow the rules of scientific integrity. At the same time tell them that acting against the rules on testing, will violate that trust [6].

5 Pilots

Pilots were run to test the solution with students [14]. Two pilots were run in April 2016 in two exams at the Eindhoven University of Technology. The characteristics of the courses were chosen as diverse as possible. One course was from the Electro Engineering department dealing with embedded systems design. The other course was dealing with Academic Writing in English, and could be followed by all students.

For both pilots, a Moodle environment (i.e., OnCourse) was used as online test suite. The Electro Engineering assessment comprised multiple choice questions and short answer questions. The English course assessed the students by using two open questions with an open text field for the students to write their answers.

In total, 135 students were assessed. 102 students (76%) used a windows notebook from the TU/e notebook arrangement. 33 students used another notebook. All students (100%) were able to make and successfully send in all their answers of the test. The number of successful test administration using the USB-stick was 104 (77%). The others (23%) did not use the memory stick to boot their notebook but went to the test environment by using their normal browser. Extra monitoring by proctors secured these exams. The reason for not being able to use the USB-stick was diverse (e.g., students not willing to use the solution, exotic notebooks that did not boot from the stick, and problems with Macs). At the time of this pilot the main reason for not being able to boot from the memory stick was a sudden problem with almost all Macs.

From the evaluation it was learned that overall students perceive the digital assessment with the USB-stick as all right but that they are worried about a situation when technology fails. Teachers are happy to use digital assessment as the correction work is easier and faster.

5.1 Conclusions and Recommendations from the Pilot Setup for Further Development

The proof of concept that was used in the pilot is described in detail by Goossens and Koedam [10]. Here we sum the main advantages and disadvantages. Starting with the first:

- The USB-stick solution by Goossens and Koedam [8] is in line with analysis above. Control is taken over early in the booting process of the students notebooks. Possibilities for students to 'prepare' their notebook for cheating are limited (i.e., in institution control)
- The USB-stick itself is a physical token with a unique hardware ID. This allows the use of black/white listing of these tokens as part of the STEP security measures.
- The USB-stick solution allows the use of Wi-Fi so the establishment of test halls with wired connections is not necessary. The secured environment only allows the network card to connect with the safe test server (or other trusted environments where software can be run).

Some disadvantages to deal with are:

- There is no guarantee that all notebooks can boot from the USB-stick. However, this can be known on forehand, tested and possibly be solved per occasion.
- The notebook market should continuously be monitored for changes, to avoid (groups of) notebooks that suddenly do not work. A practical implementation is to check all

student notebooks at the start of the academic year. But still sudden changes in the BIOS could have impact on the performance.
- Software (e.g., Matlab, SPSS, Word) are not available yet on USB stick due to several challenges (e.g., licenses, suitable for Linux, stick size)

The main recommendations for the STEP solution, based on the pilots are focused at the further development of the solution. For further development it is important firstly to arrange the ownership of the solution in the organization to insure a fast reaction in case of security problems and/or problems with functionality (e.g., sudden failure of the solution on all Macs). Then there is the need to improve the procedures prior and during the assessment and communicate the procedures to all stakeholders (e.g., students, proctors) and focus at the teachers' needs for essential software in assessment (e.g., SPSS, Matlab),

6 Preliminary Conclusion and Discussion

Based on the reports [8, 10] and the experience form the pilots [14], the estimation is that the use of personal devices (i.e., notebooks) in taking high stake tests is possible (as inner shell of a complete STEP solution). The core of the solution is a (Linux) boot from a USB memory allowing Wi-Fi connection to the test server only. However, for the time being, the solution only seems to work in a situation with a homogeneous selection of devices (i.e., notebooks) including a need for an individual test of students notebooks prior to their first exam. In addition, there is the continuous need to prevent and to detect the use of virtual machines. A secured test environment should be created (i.e., STEP, shell 2) where attempts of fraud can be detected. Last but not least, the organization should arrange a third shell (i.e., shell 3), that arranges the passage of cases of fraud and arranges the reaction in a more technical way when fraud and a leak in shell 1 and 2 have been detected. This shell should also comprise an owner of the system who is responsible for the continuous testing procedures that are necessary to prevent a sudden failure of (groups of) notebooks that should work with the USB boot.

Acknowledgement. We wish to thank all colleagues and students who made it possible to run the pilots. In particular thank to prof. dr. Kees G.W. Goossens, ir. Martijn L.P.J. Koedam, Rob C. F. van den Heuvel, and ir. PDEng Huub M.P. de Hesselle for sharing their knowledge.

References

1. SURF: Richtsnoer veilige digitale toetsafname [Guidebook secure digital assessment]. SURF, Utrecht, The Netherlands (2014)
2. SURF: Werkboek veilig toetsen [Workbook secure assessment]. SURF, Utrecht, The Netherlands (2016)
3. Bloothoofd, G., Hoiting, W., Russel, K.: Plagiaatbeleid aan de Universiteit Utrecht, faculteiten en opleidingen [Plagiarism prevention at the Utrecht University, departments and programs]. Utrecht, the Netherlands (2004)

4. Duggan, F.: Plagiarism: prevention, practice and policy. Assess. Eval. High. Educ. **31**, 151–154 (2006). doi:10.1080/02602930500262452
5. Park, C.: In other (people's) words: plagiarism by university students - literature and lessons. Assess. Eval. High. Educ. **28**, 471–488 (2003). doi:10.1080/02602930301677
6. Van Meeuwen, L.W., Kraak, I.T.C.: TU/e Educational Fraud Policy. Eindhoven University of Technology, The Netherlands (2015)
7. Rienties, B., Arts, M.: Omgaan met plagiaat: Van intuïtie naar bewijs [Dealing with plagiarism: From intuition towards evidence]. Tijdschrift voor Hoger Onderwijs **4**, 251–264 (2004)
8. Verhagen, F.: Mogelijkheden gecontroleerd digitaal toetsen [Possibilities secured digital assessment]. Eindhoven University of Technology, The Netherlands (2015)
9. SURF. Digitaal toetsen, ga er maar aan staan [Digital assessment, the challenges]. Utrecht, The Netherlands, December 2015. doi:https://www.surf.nl/agenda/2015/12/tweedaags-seminar-digitaal-toetsen-ga-er-maar-aan-staan/index.html
10. Goossens, K.G.W., Koedam, M.L.P.J.: Fraud-resistant computerised examinations. ES Reports, Eindhoven University of Technology, The Netherlands (2016). doi:http://www.es.ele.tue.nl/esreports/esr-2015-03.pdf
11. LaPiedra, J.: The information security process prevention, detection and response. Global Inf. Assur. Cert. Paper, 1–6 (2000). DOI:https://www.giac.org/paper/gsec/501/information-security-process-prevention-detection-response/101197
12. Schneider, B.: The future of incident response. IEEE Comput. Reliab. Soc., 95–96 (2014). doi:https://www.schneier.com/blog/archives/2014/11/the_future_of_i.html
13. Verhagen, F.P., De Hesselle, H.M.P.: Risicoanalyse digitaal toetsen via BYOD [Risk analysis digital assessment in BYOD]. Eindhoven University of Technology, The Netherlands (2016)
14. Van den Heuvel, R., Van Meeuwen, L.W., Verhagen, F.P.: Pilot beveiligd toetsen TU/e [Pilot secured assessment TU/e]. Working group pilot 'secured assessment'. Eindhoven University of Technology, The Netherlands (2016)

What Types of Essay Feedback Influence Implementation: Structure Alone or Structure and Content?

Denise Whitelock[1(✉)], Alison Twiner[1], John T.E. Richardson[1], Debora Field[2], and Stephen Pulman[2]

[1] The Open University, Milton Keynes, UK
{denise.whitelock,alison.twiner,john.t.e.richardson}@open.ac.uk
[2] University of Oxford, Oxford, UK
{debora.field,stephen.pulman}@cs.ox.ac.uk

Abstract. Students have varying levels of experience and understanding, and need support to inform them of expectations and guide their learning efforts. Feedback is critical in this process. This study focused on the effects of providing different types of feedback on participants' written essays and on participants' motivations for learning using measures of motivation and self-efficacy. We examined whether participants performed differently in subsequent essays after receiving feedback on structure alone or on structure and content; whether their self-reported levels of motivation and attitudes to learning were related to essay performance; and whether the difference in type of feedback affected their self-reported levels of motivation and attitudes to learning. Findings revealed no significant difference in marks between those receiving feedback on structure alone and those receiving feedback on structure and content. Even so, using feedback to highlight structural elements of essay writing can have a positive impact on future essay performance.

Keywords: Essay structure · Feedback · Motivation · Self-efficacy · Self-reports

1 Introduction

People come to educational courses and learning tasks with varying levels of experience and understanding. They therefore need support in their courses to inform them of expectations and to guide their learning efforts. They also need feedback on their performance, so that they can gauge how their performance aligns with expectations, and how they can improve their efforts and attainments. This is particularly important when supporting distance and online learners, who may have little or no face-to-face contact with instructors or peers. This paper focuses on the effects of providing different types of feedback electronically on participants' written essays, as well as participants' motivations for learning using measures of motivation and self-efficacy. In terms of the research questions, it was important to ascertain whether participants performed differently in subsequent essays after receiving feedback on structure alone, or structure and content. The research team also set out to investigate whether this difference in type of feedback affected participants' self-reported levels on a standardised questionnaire regarding motivation and attitudes to learning, and whether their self-reported levels on these measures were related to their essay performance.

Such a study regarding feedback and performance linked to motivations for learning and self-efficacy has considerable implications for supporting students to improve their work, and also supporting students to *believe* that they can improve their academic work—no small feat for learners who may often feel isolated and stretched trying to squeeze study around other commitments and demands on their time. This work is intended to offer some illumination on the kinds of feedback given to written academic essays that students pay attention to, find useful and ultimately implement in future writing efforts, and how this interacts with their motivation and self-efficacy. Such findings could be of substantial benefit to students and instructors alike. This paper begins by reviewing how the literature portrays the key issues of feedback, including aspects related to motivation and self-efficacy.

1.1 Feedback

Feedback has been a popular topic of educational research for some decades, and it is largely agreed that feedback is central to learning [1]. Some researchers have however argued that the positive effects of feedback are not guaranteed [2], and so it is important that research continues to investigate how feedback can be offered in ways that support improvements in students' learning (understanding of topics) as well as performance (marks achieved). Chickering and Gamson [3] listed "gives prompt feedback" and "encourages active learning" as two of their seven principles of good practice for undergraduate education. Therefore by this argument facilitating students to take ownership of and reflect on their work, through provision of feedback at a point where they can build on it in subsequent tasks, should have the most impact on students' understanding of the requirements of academic writing.

Nelson and Schunn [4] carried out a correlational analysis of 1,073 segments of peer review feedback that had been given to undergraduate students on writing tasks. In terms of making use of feedback, Nelson and Schunn proposed that understanding problems raised within feedback about one's own work was a critical factor in implementing suggestions. They continued to explore this potential, in stating that understanding was more likely where those giving feedback provided solutions, specified where identified problems occurred, and summarised performance. Nelson and Schunn identified feedback as involving motivation, reinforcement, and information. They addressed five features of feedback: summarisation; specificity; explanations; scope (i.e., local versus global); and affective language (praise, inflammatory, and mitigating language). The first four are cognitive features, whereas the latter is an affective feature. It is these five features of feedback that were drawn on in the present study in determining the types of feedback to offer participants on their draft essays.

Nelson and Schunn proposed that there are "mediators" that operate between the provision of feedback features and implementation of suggestions. The authors addressed the mediators of "understanding feedback" and "agreement with feedback". They suggested that cognitive feedback features were most likely to influence understanding, but that affective features were more likely to influence agreement and hence implementation. Nelson and Schunn's results therefore showed how understanding feedback is critical to implementing suggestions from feedback. Thus, it is important in course design that consideration is given to how to increase the likelihood that feedback is understood, if students are to make use of it in current and future work—to learn from it (and improve performance) by understanding it, rather than just improving one-off performance by blind implementation.

Lee [5] argued that feedback research needed to address real-life writing tasks and contexts in order to have pedagogical value. Hawe and Dixon [6] adopted a case study approach to investigate "best practice" regarding feedback among teachers in New Zealand. They concluded that students needed to be given authentic opportunities to assess and edit their own work during its production to gain a better understanding of what constitutes a "good" piece of writing. This would increase the likelihood of them being able implement suggestions made in feedback and to learn from the feedback given both to themselves and to other students. Baker [7] drew attention to the distinction between "feedback" (to be used to inform future writing) and ("grading" as a final evaluative mark allocated to work already done). Even so, she reported that in practice the terms were often used interchangeably. Baker had adopted an ethnographic approach at one U.S. university and provided an in-depth analysis of teachers' views and approaches to giving feedback. However, she said little about what was actually covered in the feedback (such as content, structure, or grammar).

The latter issue has been discussed specifically with regard to the learning of English as a second language (ESL). In secondary education in Hong Kong, Lee [8] found that, in spite of receiving feedback, students were concerned that they would make the same errors again, indicating that in this context the main focus of feedback was on error correction rather than on content or organisation. Lee suggested that students might not be so error-focused if feedback were focused more on content and organisation as opposed to errors. However, in subsequent research in a similar educational context, Lee [9] identified a persisting focus on error correction, resulting in an approach that emphasised "testing" rather than "teaching". Hyland [10] had similarly observed the predominance in ESL practices of feedback as "evaluation" rather than "education". Ferris, Brown, Liu, and Stine [11] argued that it was important to note the similarities between the feedback given to students writing in English as their first language and the feedback given to those writing in English as a second language. Such considerations led Hyland [12] to argue that we must address students' response to feedback as well as the feedback itself, which is very much the focus of the present study.

Relevant to the idea that feedback should be aimed at "education" rather than solely "evaluation", Wang and Wu [13] considered feedback in terms of knowledge of results, knowledge of correct response, and elaborative feedback. They commented:

Research suggests that feedback is one of the most significant sources of information helping individual students to correct misconceptions, reconstruct knowledge, support metacognitive processes, improve academic achievement, and enhance motivation (p. 1591). (Clark & Dwyer, 1998; Foote, 1999; Warden, 2000; Zimmerman & Martinez-Pons, 1992)

From this it is apparent that feedback can play a key role in enhancing students' motivation to study and to improve their work. Therefore it is appropriate now to turn attention more specifically to issues of motivation.

1.2 Motivation

Schraw [14] proposed that self-regulated learning (SRL) consisted of knowledge, metacognition, and motivation. Banyard, Underwood, and Twiner [15] also outlined that, whilst SRL refers to learners' goals and knowledge of strategies to work towards those goals, it is highly susceptible to external influences. So how can external influences be presented to facilitate students' SRL efforts? In the current study, the aim was to guide the external influences perceived by participants by giving feedback that would support them in considering the requirements of academic writing in the context of their own work and suggest where there might be areas for development. The onus was therefore on the participants to incorporate this feedback within their subsequent essay writing. Related to motivation is the concept of self-efficacy, which will now be addressed.

1.3 Self-efficacy

Self-efficacy refers to students' judgments of how well they can do a task [16]. Much research on self-efficacy refers back to Bandura's original definition and work [17–19]. García Duncan and McKeachie [20] offered a definition of self-efficacy that incorporated "expectancy for success (which is specific to task performance) and judgments of one's ability to accomplish a task and confidence in one's skills to perform a task" (p. 119). In a survey of the use of electronic information resources among university students in Nigeria, Tella, Tella, Ayeni, and Omoba [21] also defined self-efficacy as whether an individual feels capable of doing a particular task or succeeding in a current situation. In other words, self-efficacy is a matter of the perception of one's own ability in a perceived context, rather than a measure of ability itself. Tella et al. concluded:

> The results indicate that self-efficacy and the use of electronic information jointly predict and contribute to academic performance; that respondents with high self-efficacy make better use of electronic information and have better academic performance; that a correlation exists among self-efficacy, use of electronic information and academic performance (Abstract).

Thus, raising students' self-efficacy is an important aim. Tang and Tseng [18] carried out a survey of online distance learners and reported:

> Distance learners who have higher self-efficacy for information seeking and proficiency in information manipulation exhibited higher self-efficacy for online learning. Moreover, students with high self-efficacy demonstrated superior knowledge of digital resources selection. Students who have low self-efficacy with regard to information seeking were more likely to express interest in learning how to use the library resources, although learning techniques for database searching was the exception (p. 517).

Such findings have strong implications for the support that can be offered to enable students to increase their self-efficacy, by offering opportunities for them to review, reflect on and improve their work and so build on progressive successes, rather than receiving only final-grade evaluations with little scope to learn from mistakes. This is particularly important in distance learning contexts, where there is minimal or no opportunity for face-to-face interaction between tutors and students or among students themselves.

Shrestha and Coffin [22] reported on a study to support distance learning students' academic writing exercises. Instructors on the course provided support and feedback to students via email, in response to successive drafts of assignments. However, the study and the detailed feedback interactions described by Shrestha and Coffin only involved two students. It is likely, therefore, that this was a very valuable provision, but one that would be difficult to resource on a large scale.

In stark contrast in terms of research design, Wang, Shannon, and Ross [23] conducted structural equation modelling of survey data from 256 graduate and undergraduate students taking online courses with one U.S. university. They reported a series of effects:

> Students with previous online learning experiences tended to have more effective learning strategies when taking online courses, and hence, had higher levels of motivation in their online courses. In addition, when students had higher levels of motivation in their online courses, their levels of technology self-efficacy and course satisfaction increased. Finally, students with higher levels of technology self-efficacy and course satisfaction also earned better final grades (p. 302).

This again shows a finding of a cumulative effect for online distance learners, in terms of students using effective learning strategies, having high levels of motivation, self-efficacy and course satisfaction, and achieving better final grades. Much of this therefore comes down to the individual learner's stance toward their study. However, much can also be done by instructors and course designers to support students in selecting appropriate learning strategies and understanding task requirements and in enabling students to feel that they can do well. It is this aim for feedback that is addressed in the study reported here.

1.4 Research Questions and Analyses

In the current study, participants completed adapted versions of the Motivated Strategies for Learning Questionnaire (MSLQ) [24, 25] before and after writing two essays. They were given qualitative feedback (based on the marks awarded but not including the marks awarded) after submitting each essay. One group of participants received feedback only on the structure of their essays, while the second group received feedback on both the structure and the content of their essays.

The first research question was:

- Does the kind of feedback that the participants receive for the first essay influence the marks that they obtain on the second essay?

The fact that the marks awarded for the second essay might be different from the marks awarded for the first essay is in itself not surprising, because the essays were written on different topics (although the marking criteria were not topic-specific). This question asks whether the difference was greater for the participants who received feedback on the structure and the content of their essays than for the participants who only received feedback on the structure of their essays. The latter would entail an interaction between the effect of groups and the effect of essays (first versus second) on the marks awarded.

The second research question was:

- Do the scores the participants obtain on the pre-test questionnaire predict the marks that they receive for their essays?

An additional question here is:

- Is the regression between the questionnaire scores and the essay marks the same regardless of whether the participants receive feedback only on the structure of their essays or feedback on both the structure and content of their essays?

Preliminary tests were carried out to check for interactions between the pre-test questionnaire scores and the effect of groups on the marks awarded to test for the homogeneity of the regression between the pre-test questionnaire scores and the essay marks.

The third research question was:

- Do the scores that participants obtain on the post-test questionnaire differ, depending on the kind of feedback that they receive for their essays?

The final research question was:

- Do the marks that participants receive for their essays predict the scores that they obtain on the post-test questionnaire?

A complication is that, if the marks depended on their scores on the pre-test questionnaire, the marks might simply serve as a mediator between the pre-test scores and the post-test scores. This point can be addressed by asking whether the marks awarded for the essays predicted their post-test scores when the effects of the pre-test scores on the latter had been statistically controlled. Once again, an additional question is:

- Is the regression between the marks awarded for the essays and the scores on the post-test questionnaire the same regardless of whether the participants received feedback only on the structure of their essays or feedback on both the structure and content?

A preliminary test was carried out to check for an interaction between the essay marks and the effect of groups on the post-test questionnaire scores to test for the homogeneity of the regression between the essay marks and the post-test questionnaire scores. Similar tests were carried out to check for interactions between the pre-test questionnaire scores and the effect of groups on the post-test scores to test for the homogeneity of the regression between the pre-test scores and the post-test scores.

2 Method

2.1 Participants

Ninety-one participants were recruited from a subject panel maintained by colleagues in the Department of Psychology consisting of people who were interested in participating in online psychology experiments. Some of them were current or former students of the Open University, but others were just members of the public with an interest in psychological research.

2.2 Materials

The MSLQ consists of 81 statements in 15 scales that measure motivational variables and learning strategies on particular courses. Respondents are asked to rate each statement on a 7-point scale from 1 for "not at all true of me" to 7 for "very true of me". Pintrich et al. [24] commented that the 15 scales were "designed to be modular" so that they "can be used together or singly" (p. 3). The present investigation used items drawn from the first six scales, which are concerned with respondents' motives and attitudes. Pintrich et al. [25] explained these scales as follows:

> The motivational scales are based on a general social-cognitive model of motivation that proposes three general motivational constructs: (1) expectancy, (2) value, and (3) affect. Expectancy components refer to students' beliefs that they can accomplish a task. Two expectancy-related subscales were constructed to assess students' (a) perceptions of self-efficacy and (b) control beliefs for learning. Value components focus on the reasons why students engage in an academic task. Three subscales are included in the MSLQ to measure value beliefs: (1) intrinsic goal orientation (a focus on learning and mastery), (2) extrinsic goal orientation (a focus on marks and approval from others), and (3) task value beliefs (judgments of how interesting, useful, and important the course content is to the student). The third general motivational construct is affect, and has been operationalised in terms of responses to the test anxiety scale, which taps into students' worry and concern over taking exams (p. 802).

The MSLQ was intended for use with students taking formal courses. For the present investigation, references to "this class" or "this course" were reworded to refer to "this exercise". One item from the Intrinsic Goal Orientation scale and four items from the Test Anxiety scale could not be adapted, leaving 26 items. These were phrased in the present or future tense in the pre-test questionnaire and in the present or past tense for the post-test questionnaire. Sample items are shown in Table 1. Both surveys were hosted online on secure websites.

2.3 Procedure

Communication with the participants was solely online. They were assigned alternately to two groups. (Those in Group 1 were provided with feedback only on the structure of their essays; those in Group 2 were provided with feedback on both the structure and content of their essays.) Each participant was asked to complete the pre-test version of the MSLQ online, for which they were allowed 2 weeks. They were then asked to "write an essay on human perception of risk" of between 500 and 1,000 words, for which they were also allowed 2 weeks.

Two of the authors who were academic staff with considerable experience in teaching and assessment marked the submitted essays using an agreed marking scheme and without reference to the groups to which participants had been assigned. The marking scheme is shown in Table 2. If the difference between the total marks awarded was 20% points or less, essays were assigned the average of the two markers' marks. Discrepancies of more than 20% points were resolved by discussion between the markers.

The participants were provided with qualitative feedback on their first essays based on the marks awarded (but they were not given the marks themselves). Those in Group 1 received feedback based on the marks awarded against Criteria 1, 2, 5, 6, and 7. Those in Group 2 received feedback based on the marks awarded against all 10 criteria. They were then asked to "write an essay on memory problems in old age" of between 500 and 1,000 words, for which they were again allowed 2 weeks. Finally, they were provided with feedback on their second essays and asked to complete the post-test version of the MSLQ online. Participants who completed both questionnaires and submitted both essays were rewarded with an honorarium of £40 sterling in Amazon vouchers.

3 Results

Of the 91 participants who were invited to complete the pre-test questionnaire, 76 responded, but two only provided incomplete data. Of these 76 participants, 42 submitted Essay 1, of whom 38 submitted Essay 2. Of the latter 38 participants, all completed the post-test questionnaire. They consisted of eight men and 30 women; they were aged between 23 and 65 (mean age = 42.3 years); and 19 participants were in each group.

3.1 Marks Awarded for Essays

The correlation coefficients between the marks initially awarded by the two markers were .85 for Essay 1 and .82 for Essay 2. The discrepancy between the two markers was more than 20% points for just one essay, and this discrepancy was resolved by discussion between the markers. The mean final mark for Essay 1 was 56.8 (SD = 14.7), and the mean final mark for Essay 2 was 63.3 (SD = 12.4).

A mixed-design analysis of variance was carried out on the average mark that was awarded to the participants who submitted two essays. This employed the within-subjects variables of essays (Essay 1 versus Essay 2) and marking criteria (1–10) and the between-subjects variables of group (Group 1 versus Group 2). Post hoc tests were carried out to identify the marking criteria on which any significant differences in marks had arisen.

There was no significant difference in the overall marks awarded to participants in the two groups, $F(1, 36) = 2.83$, $p = .10$, partial $\eta^2 = .07$. The mean mark awarded for

Table 1. Changes of wording in sample MSLQ items for pre-test and post-test questionnaires (original wording quoted from Pintrich et al. [24]).

Intrinsic Goal Orientation scale

Original wording: In a class like this, I prefer course material that really challenges me so I can learn new things.

Pre-test questionnaire: In an exercise like this, I prefer tasks that really challenge me.

Post-test questionnaire: In an exercise like this, I prefer tasks that really challenge me.

Extrinsic Goal Orientation scale

Original wording: Getting a good grade in this class is the most satisfying thing for me right now.

Pre-test questionnaire: Getting a good grade in this exercise is the most satisfying thing for me.

Post-test questionnaire: Getting a good grade in this exercise was the most satisfying thing for me.

Task Value scale

Original wording: I think I will be able to use what I learn in this course in other courses.

Pre-test questionnaire: I think I will be able to use what I learn in this exercise in other situations.

Post-test questionnaire: I think I will be able to use what I learned in this exercise in other situations.

Control of Learning Beliefs scale

Original wording: If I study in appropriate ways, then I will be able to learn the material in this course.

Pre-test questionnaire: I will be able to master the material needed for this exercise.

Post-test questionnaire: I was able to master the material needed for this exercise.

Self-Efficacy for Learning and Performance scale

Original wording: I believe I will receive an excellent grade in this class.

Pre-test questionnaire: I believe I will receive an excellent grade for this exercise.

Post-test questionnaire: I believe I received an excellent grade for this exercise.

Test Anxiety scale

Original wording: When I take a test I think about how poorly I am doing compared with other students.

Pre-test questionnaire: I will think about how poorly I am doing compared with the other participants.

Post-test questionnaire: I thought about how poorly I was doing compared with the other participants.

Table 2. Marking scheme for essays.

Criterion	Maximum marks
1. Introductory paragraph sets out argument	10
2. Concluding paragraph rounds off discussion	10
3. Argument is clear and well followed through	10
4. Evidence for argument in main body of text	20
5. All paragraphs seven sentences long or less	5
6. Word count between 500 and 1,000 words	5
7. Award 5 for two or three references, 10 for four or more	10
8. Provides a clear and explicit definition of risk or memory	10
9. Extensive vocabulary, accurate grammar and spelling	10
10. Understanding of practical issues, innovative proposals	10

Essay 2 (63.1) was significantly higher than the mean mark awarded for Essay 1 (57.3), $F(1, 36) = 4.89$, $p = .03$, partial $\eta^2 = .12$. However, the interaction between the effects of groups and essays was not significant, $F(1, 36) = .40$, $p = .53$, partial $\eta^2 = .01$, implying that the difference between the marks awarded for the two essays was similar for participants in the two groups.

The main effect of criteria was statistically significant, $F(9, 324) = 39.70$, $p < .001$, partial $\eta^2 = .52$, which is unsurprising since different numbers of marks were awarded against the 10 criteria. There was a significant interaction between the effect of groups and the effect of criteria, $F(9, 324) = 2.50$, $p = .01$, partial $\eta^2 = .07$, implying that the two groups showed a different pattern of marks across the different criteria. There was also a significant interaction between the effect of essays and the effect of criteria, $F(9, 324) = 6.69$, $p < .001$, partial $\eta^2 = .16$, implying that the two essays showed a different pattern of marks across the different criteria. Finally, there was a significant three-way interaction between the effects of groups, essays and criteria, $F(9, 324) = 2.05$, $p = .03$, partial $\eta^2 = .05$.

Post hoc tests showed that the difference between the marks awarded to the two groups was only significant on Criterion 8 (Definition), where Group 2 obtained a higher mean mark (5.70) than did Group 1 (3.21), $F(1, 36) = 7.24$, $p = .01$, partial $\eta^2 = .17$. However, this was qualified by a significant interaction between the effect of groups and the effect of essays, $F(1, 36) = 5.12$, $p = .03$, partial $\eta^2 = .13$. For the participants in Group 1, the mean mark awarded on Criterion 8 was higher on Essay 2 (4.03) than on Essay 1 (2.40). However, for the participants in Group 2, the mean mark awarded on Criterion 8 was higher on Essay 1 (6.50) than on Essay 2 (4.90). In short, there was no evidence that providing feedback on both the structure and content of an essay led to higher marks on a subsequent essay than providing feedback on the first essay's structure alone.

Finally, the difference between the marks awarded to the two essays was significant on Criterion 4, $F(1, 36) = 12.86$, $p = .001$, partial $\eta^2 = .26$, on Criterion 7, $F(1, 36) = 19.82$, $p < .001$, partial $\eta^2 = .36$, and on Criterion 9, $F(1, 36) = 5.94$, $p = .02$, partial $\eta^2 = .14$. The mean mark awarded on Criterion 4 (Evidence) was higher for

Essay 2 (11.12) than for Essay 1 (8.97). The mean mark awarded on Criterion 7 (References) was also higher for Essay 2 (7.22) than for Essay 1 (4.01). However, the mean mark awarded on Criterion 9 (Written presentation) was higher for Essay 1 (7.93) than for Essay 2 (7.33).

3.2 Questionnaire Scores

The participants were assigned scores on the pre-test and post-test questionnaires by calculating the mean response to the constituent items in each scale. Table 3 shows the means and standard deviations of these scores, together with the relevant values of Cronbach's [26] coefficient alpha as an estimate of reliability. The latter were broadly satisfactory on conventional research-based criteria [27].

Table 3. Means, standard deviations, and values of coefficient alpha on pre-test and post-test questionnaires.

Scale	No. of items	Pre-test questionnaire			Post-test questionnaire		
		M	SD	Alpha	M	SD	Alpha
Intrinsic goal orientation	3	5.55	.97	.70	5.50	0.89	.65
Extrinsic goal orientation	4	4.32	1.19	.78	4.40	1.17	.74
Task value	6	5.16	.81	.79	5.41	0.94	.87
Control of learning beliefs	4	4.97	1.07	.75	5.13	0.98	.64
Self-efficacy	8	4.78	.91	.92	5.01	0.91	.88
test anxiety	1	3.37	1.73	–[a]	2.87	1.71	–[a]

[a]Coefficient alpha cannot be calculated for scales consisting of a single item.

The correlation coefficients between corresponding scales in the pre-test questionnaire and the post-test questionnaire were highly significant and ranged between +.46 and +.74, except in the case of Extrinsic Goal Orientation, $r = +.31$, $p = .06$.

A doubly multivariate analysis of variance was carried out on the scores that the participants obtained on the pre-test and post-test questionnaires. This used the between-subjects variable of groups and the within-subject comparison between the pre-test and post-test scores. This analysis found that on Self-Efficacy the post-test scores were significantly higher than the pre-test scores, $F(1, 36) = 4.19$, $p = .04$, partial $\eta^2 = .10$. However, there were no other significant differences between the pre-test scores and the post-test scores, no significant differences between the scores obtained by the two groups, and no significant interactions between these two effects. Thus, the two groups were similar in terms of their MSLQ scores both at the beginning and at the end of the study, and the increase in scores on Self-Efficacy occurred regardless of the kind of feedback they had received.

3.3 Using the Pre-test Scores to Predict the Essay Marks

A univariate analysis of variance was carried out using the between-subjects variable of group, the scores on the pre-test questionnaire as covariates, and the average essay mark as the dependent variable. A preliminary analysis included the interactions between the scores on the pre-test questionnaire and the effect of group to check for homogeneity of regression. This found no significant interactions between the scores on the pre-test questionnaire and the effect of groups on the average essay mark, $F(1, 24) \leq 1.47$, $p \geq .24$, partial $\eta^2 \leq .06$, implying homogeneity of regression between the two groups. This analysis also found no significant difference between the two groups in terms of their average essay mark when their pre-test scores were taken into account, $F(1, 24) = .84$, $p = .37$, partial $\eta^2 = .03$. Accordingly, the difference between the two groups was ignored in the main analysis.

This used the scores on the pre-test questionnaire as covariates and the average essay mark as the dependent variable. The average essay mark was significantly predicted by the scores on the Task Value scale, $B = +6.93$, $F(1, 31) = 5.61$, $p = .02$, partial $\eta^2 = .15$, and by the scores on the Control of Learning Beliefs scale, $B = +4.37$, $F(1, 31) = 4.21$, $p = .04$, partial $\eta^2 = .12$, but not by the scores on any of the other four scales. Thus, those participants who produced higher scores on Task Value and Control of Learning Beliefs in the pre-test questionnaire tended to obtain higher marks for their essays.

3.4 Using the Pre-test Scores and the Essay Marks to Predict the Post-test Scores

A multivariate analysis of variance was carried out using the between-subjects variable of group, the scores on the pre-test questionnaire, and the average essay mark as co-variates and the scores on the post-test questionnaire as dependent variables. A preliminary analysis included the interactions between the scores on the pre-test questionnaire and the effect of group and the interaction between the average essay mark and the effect of group to check for homogeneity of regression. This found that there was a significant interaction between the scores on Test Anxiety on the pre-test questionnaire and the effect of groups on the scores on the post-test questionnaire, $F(6, 17) = 6.60$, $p = .001$, partial $\eta^2 = .70$. In particular, this interaction was significant for the scores on there was a significant interaction between the scores on Test Anxiety on the pre-test questionnaire and the effect of groups on the scores on the post-test questionnaire, $F(6, 17) = 6.60$, $p = .001$, partial $\eta^2 = .70$. In particular, this interaction was significant for the scores on Control of Learning Beliefs in the post-test questionnaire, $F(1, 22) = 26.78$, $p < .001$, partial $\eta^2 = .55$, and the scores on Self-Efficacy on the post-test questionnaire, $F(1, 22) = 5.00$, $p = .04$, partial $\eta^2 = .19$, but not on the other four scales of the post-test questionnaire. Separate analyses carried out on the two groups showed that the pre-test scores on Test Anxiety were negatively correlated with the post-test scores on Control of Learning Beliefs in Group 1, $B = -.36$, $F(1, 12) = 11.73$, $p = .005$, partial $\eta^2 = .49$, but were positively correlated with the post-test scores on Control of Learning Beliefs in Group 2, $B = +.65$, $F(1, 12) = 16.93$,

$p = .001$, partial $\eta^2 = .59$. A similar pattern was evident in the post-test scores on Self-Efficacy, but neither of the groups demonstrated a significant regression coefficient in this case. None of the other interaction terms was significant in the preliminary analysis, implying homogeneity of regression between the two groups in other respects, including the regression between the average essay mark and the post-test scores. The latter interaction terms were dropped from the main analysis.

This used the between-subjects variable of group, the scores on the pre-test questionnaire and the average essay mark as covariates and the scores on the post-test questionnaire as dependent variables. The statistical model included the interaction between the scores on Test Anxiety on the pre-test questionnaire and the effect of groups. This remained significant for the scores on Control of Learning Beliefs in the post-test questionnaire, $F(1, 28) = 24.38$, $p < .001$, partial $\eta^2 = .47$. The mean score obtained by Group 1 on this scale (5.30) was significantly higher than the mean scores obtained by Group 2 (5.03), $F(1, 28) = 24.53$, $p < .001$, partial $\eta^2 = .47$. Finally, there was a positive relationship between the participants' average essay mark and their scores on Extrinsic Goal Orientation in the post-test questionnaire, $B = +.06$, $F(1, 28) = 6.93$, $p = .01$, partial $\eta^2 = .20$. In other words, even when the participants' scores on the pre-test questionnaire had been statistically controlled, those who obtained higher marks for their essays subsequently tended to produce higher scores on Extrinsic Goal Orientation in the post-test questionnaire.

4 Discussion

This paper has reported on a study addressing the effects of offering feedback regarding structure or regarding structure and content on written academic essays. The study also focused on participants' reported levels of self-efficacy and motivation for such learning tasks, using an adapted version of the MSLQ. Our intention was that feedback given on a first essay would be implemented and supportive in writing a second essay. We also hypothesised that there would be a difference in marks related to the type of feedback participants received—on structure or on structure and content—as well as a relationship between marks received and reported levels of motivation and self-efficacy.

Based on the existing literature, there can be reciprocal relationships between motivation, self-efficacy, and improving academic performance, whereby enhancements in the former often correspond with improvements in the latter [21]. The current study therefore uniquely set out to measure aspects of motivation and self-efficacy before and after the study activities—writing two essays on which participants were given qualitative feedback (but not numerical marks)—and to see whether providing different types of feedback affected participants' reported levels of motivation and self-efficacy, as well as the essay marks they achieved. Analysis of the collected data revealed a number of interesting findings.

First, there was no evidence that providing feedback on both structure and content, compared to just structure, led to a higher mark being achieved on the second essay. For both groups however, marks allocated for the criteria regarding use of evidence and references significantly increased on Essay 2, suggesting that feedback received on Essay 1 may have encouraged participants to focus on these aspects in their second essay.

With regard to the participants' MSLQ responses, both groups had similar scores for their pre-test and post-test questionnaires, apart from values on the self-efficacy scale, which showed significantly higher scores on the post-test regardless of group. This suggests that all participants increased their self-efficacy by the end of the study, regardless of the type of feedback received.

Linking the pre-test scores and essay marks, no evidence was found of an effect of the type of feedback received. It was apparent, however, that those participants who scored more highly on the task value and control of learning beliefs scales in their pre-test questionnaires also tended to achieve higher marks in their essays. Linking pre-test scores, essay marks and post-test scores, evidence was found that those who scored more highly on their essays also reported having higher extrinsic goal orientation on the post-test questionnaire.

5 Conclusion and Implications

Together these findings tell a complicated but interesting story. For instance, participants' marks on their second essays were similar whether they received feedback on essay structure alone or on both structure and content. However, both groups received higher marks on their second essays with regard to use of evidence and references (key aspects of essay structure highlighted in the feedback) after receiving feedback on their first essay. This is a very important finding, as use of evidence and references are key components in writing academic essays, and the feedback provided may have supported participants in improving their efforts and subsequently their performance toward these aspects. It can only be speculated that participants' higher self-reported levels of self-efficacy in their post-test questionnaire may have been related to this improved understanding and performance. As self-efficacy is such an important factor in people feeling they can do a task they set themselves, this is a crucial indicator that such feedback may have a key role to play in helping participants to understand task requirements and how they can improve. This is supported by the finding that those participants who scored more highly on task value and control of learning beliefs in their pre-test questionnaires also scored more highly on their essays, suggesting that there was a relationship between participants who wrote better essays and those who felt they were in control of, saw the value of, and understood their efforts toward the essay-writing activities.

The lack of significant difference in marks between those receiving feedback on structure alone and those receiving feedback on structure and content is perhaps surprising and deserves further exploration. On the basis of this project it can however be concluded that using feedback to highlight certain structural elements of essay writing, in particular use of evidence and references, can have a lasting positive impact on participants' future essay performance. This is significant for all efforts to support the perception of feedback as "education" rather than just "evaluation" [9, 10], or as "advice for action" [28], as a means to help participants to improve their future work rather than simply as a mechanism for marking past work.

Acknowledgements. This work was supported by the Engineering and Physical Sciences Research Council of the United Kingdom under Grant EP/J005959/1. We wish to thank Evaghn DeSouza and Alison Green for their assistance in this project.

References

1. Black, P., Wiliam, D.: Assessment and classroom learning. Assess. Educ. **5**, 7–74 (1998)
2. Kluger, A.N., DeNisi, A.: The effects of feedback interventions on performance: a historical review, a meta-analysis, and a preliminary feedback intervention theory. Psychol. Bull. **119**, 254–284 (1996)
3. Chickering, A.W., Gamson, Z.F.: Seven principles for good practice in undergraduate education. Am. Assoc. High. Educ. Bullet. **39**(7), 3–7 (1987)
4. Nelson, M.M., Schunn, C.D.: The nature of feedback: how different types of peer feedback affect writing performance. Instr. Sci. **37**, 375–401 (2009)
5. Lee, I.: Editorial. Feedback in writing: issues and challenges. Assess. Writing **19**, 1–5 (2014)
6. Hawe, E.M., Dixon, H.R.: Building students' evaluative and productive expertise in the writing classroom. Assess. Writing **19**, 66–79 (2014)
7. Baker, N.L.: "Get it off my stack": teachers' tools for grading papers. Assess. Writing **19**, 36–50 (2014)
8. Lee, I.: Error correction in the L2 writing classroom. What do students think? TESL Canada J. **22**(2), 1–16 (2005)
9. Lee, I.: Working smarter, not working harder: revisiting teacher feedback in the L2 writing classroom. Canad. Modern Lang. Rev. **67**, 377–399 (2011)
10. Hyland, F.: Dealing with plagiarism when giving feedback. ELT J. **55**, 375–381 (2001)
11. Ferris, D., Brown, J., Liu, H.S., Stine, M.E.A.: Responding to L2 students in college writing classes: teacher perspectives. TESOL Q. **45**, 207–234 (2011)
12. Hyland, F.: Future directions in feedback on second language writing: overview and research agenda. Int. J. Engl. Stud. **10**, 171–182 (2010)
13. Wang, S.-L., Wu, P.-Y.: The role of feedback and self-efficacy on web-based learning: the social cognitive perspective. Comput. Educ. **51**, 1589–1598 (2008)
14. Schraw, G.: Measuring self-regulation in computer-based learning environments. Educ. Psychol. **45**, 258–266 (2010)
15. Banyard, P., Underwood, J., Twiner, A.: Do enhanced communication technologies inhibit or facilitate self-regulated learning? Eur. J. Educ. **41**, 473–489 (2006)
16. Bandura, A.: Perceived self-efficacy in cognitive development and functioning. Educ. Psychol. **28**, 117–148 (1993)
17. Alkharusi, H., Aldhafri, S., Alnabhani, H.: The impact of students' perceptions of assessment tasks on self-efficacy and perception of task value: a path analysis. Soc. Behav. Pers. **41**, 1681–1692 (2013)
18. Tang, Y., Tseng, H.W.: Distance learners' self-efficacy and information literacy skills. J. Acad. Librariansh. **39**, 517–521 (2013)
19. Van Dinther, M., Dochy, F., Segers, M., Braeken, J.: Student perceptions of assessment and student self-efficacy in competence-based education. Educ. Stud. **40**, 330–351 (2014)
20. García Duncan, T., McKeachie, W.J.: The making of the motivated strategies learning questionnaire. Educ. Psychol. **40**, 117–128 (2005)
21. Tella, A., Tella, A., Ayeni, C.O., Omoba, R.O.: Self-efficacy and use of electronic information as predictors of academic performance. Electron. J. Acad. Special Librariansh. **8**(2), 12 (2007) http://southernlibrarianship.icaap.org/content/v08n02/tella_a01.html

22. Shrestha, P., Coffin, C.: Dynamic assessment, tutor mediation and academic writing development. Assess. Writing **17**, 55–70 (2012)
23. Wang, C.-H., Shannon, D.M., Ross, M.E.: Students' characteristics, self-regulated learning, technology self-efficacy, and course outcomes in online learning. Distance Educ. **34**, 302–323 (2013)
24. Pintrich, P.R., Smith, D.A.F., Garcia, T., McKeachie, W.J.: A Manual for the Use of the Motivated Strategies for Learning Questionnaire (MSLQ). National Center for Research to Improve Postsecondary Teaching and Learning, University of Michigan, Ann Arbor (1991)
25. Pintrich, P.R., Smith, D.A.F., Garcia, T., McKeachie, W.J.: Reliability and predictive validity of the motivated strategies for learning questionnaire (MSLQ). Educ. Psychol. Measur. **53**, 801–813 (1993)
26. Cronbach, L.J.: Coefficient alpha and the internal structure of tests. Psychometrika **16**, 297–334 (1951)
27. Robinson, J.P., Shaver, P.R., Wrightsman, L.S.: Criteria for scale selection and evaluation. In: Robinson, J.P., Shaver, P.R., Wrightsman, L.S. (eds.) Measures of Personality and Social Psychological Attitudes, pp. 1–16. Academic Press, San Diego (1991)
28. Whitelock, D.: Activating assessment for learning: are we on the way with web 2.0? In: Lee, M.J.W., McLoughlin, C. (eds) Web 2.0-Based-E-Learning: Applying Social Informatics for Tertiary Teaching, pp. 319–342. IGI Global, Hershey (2010)

Author Index

Ackermans, Kevin 1
Anastasiou, Dimitra 11

Baudet, Alexandre 105
Berzinsh, Vsevolods 159
Brand-Gruwel, Saskia 1

De Maeyer, Sven 23
Den Brok, Perry J. 171
Djundik, Pavel 159

Field, Debora 181
Foulonneau, Muriel 105

Goossens, Maarten 23
Guerrero-Roldán, Ana-Elena 86

Hollo, Kaspar 72
Hunt, Pihel 39

Jochems, Wim 47
Joosten-ten Brinke, Desirée 47

Klinkenberg, Sharon 63

Laanpere, Mart 133
Leijen, Äli 39
Lepp, Marina 72
Luik, Piret 72

Mäeots, Mario 148
Malva, Liina 39
Mortier, Anneleen 23

Noguera, Ingrid 86
Norta, Alex 133

Palts, Tauno 72
Papli, Kaspar 72
Pedaste, Margus 148
Prank, Rein 93
Pulman, Stephen 181

Ras, Eric 11, 105
Richardson, John T.E. 181
Rienties, Bart 117
Rodríguez, M. Elena 86
Rogaten, Jekaterina 117
Rusman, Ellen 1

Saay, Salim 133
Säde, Merilin 72
Savitski, Pjotr 159
Siiman, Leo A. 148
Silm, Gerli 39
Sluijsmans, Dominique 47
Specht, Marcus 1
Suviste, Reelika 72

Tomberg, Vladimir 159
Tõnisson, Eno 72
Twiner, Alison 181

Vaherpuu, Vello 72
Van der Schaaf, Marieke 39
Van Gasse, Roos 23
Van Meeuwen, Ludo W. 171
Van Petegem, Peter 23
Vanhoof, Jan 23
Verhagen, Floris P. 171
Vlerick, Peter 23

Whitelock, Denise 117, 181

MIX
Papier aus verantwortungsvollen Quellen
Paper from responsible sources
FSC® C105338

If you have any concerns about our products,
you can contact us on
ProductSafety@springernature.com

In case Publisher is established outside the EU,
the EU authorized representative is:
**Springer Nature Customer Service Center GmbH
Europaplatz 3, 69115 Heidelberg, Germany**

Printed by Libri Plureos GmbH
in Hamburg, Germany